Computer Security Threats

Edited by Ciza Thomas, Paula Fraga-Lamas and Tiago M. Fernández-Caramés

Published in London, United Kingdom

IntechOpen

Supporting open minds since 2005

Computer Security Threats
http://dx.doi.org/10.5772/intechopen.83233
Edited by Ciza Thomas, Paula Fraga-Lamas and Tiago M. Fernández-Caramés

Contributors
Jian Zhang, Yimu Ji, Shangdong Liu, Sudha Senthilkumar, Oscar Lage, Santiago de Diego, Borja Urkizu, Eneko Gomez, Iván Gutierrez, Paula Fraga-Lamas, Tiago M. Fernández-Caramés, Ciza Thomas, M. Veena, S. Upasana, S. Prathima

First published in London, United Kingdom, 2020 by IntechOpen
IntechOpen is the global imprint of INTECHOPEN LIMITED, registered in England and Wales, registration number: 11086078, 7th floor, 10 Lower Thames Street, London, EC3R 6AF, United Kingdom
Printed in Croatia

British Library Cataloguing-in-Publication Data
A catalogue record for this book is available from the British Library

Additional hard and PDF copies can be obtained from orders@intechopen.com

Computer Security Threats
Edited by Ciza Thomas, Paula Fraga-Lamas and Tiago M. Fernández-Caramés
p. cm.
Print ISBN 978-1-83880-239-4
Online ISBN 978-1-83880-240-0
eBook (PDF) ISBN 978-1-83962-381-3

For EU product safety concerns:
IN TECH d.o.o., Prolaz Marije Krucifikse Kozulić 3, 51000 Rijeka, Croatia, info@intechopen.com or visit our website at intechopen.com.

Meet the editors

Dr. Ciza Thomas is currently working as the Director in-charge of the Directorate of Technical Education, Government of Kerala, India. Her area of expertise is network security with research interest in the fields of information security, data mining, sensor fusion, pattern recognition, information retrieval, digital signal processing, and image processing. She has more than 60 international journals and international conference proceedings and 50 national conference publications. She has edited six books and published ten book chapters. She is a reviewer of more than ten reputed international journals including IEEE transactions on Signal Processing, IEEE transactions on Neural Networks, International Journal of Network Security, International Journal of Network Management, and IEEE-John Wiley International Journal on Security and Communications Network. She is a recipient of an achievement award in 2010 and the e-learning IT award in 2014 from Government of Kerala.

Tiago M. Fernández-Caramés (Senior Member, IEEE) has worked since 2016 as an Assistant Professor in the area of Electronic Technology at the University of A Coruña (UDC) (Spain), where he obtained his MSc degree and PhD degrees in Computer Science in 2005 and 2011. Since 2005 he has worked in the Department of Computer Engineering at UDC: from 2005 to 2009 through different predoctoral scholarships and between 2007 and 2016 as Interim Professor. His current research interests include IoT/IIoT systems, RFID, wireless sensor networks, augmented reality, embedded systems and blockchain, as well as the different technologies involved in the Industry 4.0 paradigm. In such fields, he has contributed to 40 papers for conferences, to 35 articles for JCR-indexed journals and to two book chapters. Due to his expertise in the previously mentioned fields, he has acted as peer-reviewer and guest editor for different top-rank journals, and as project reviewer for national research bodies from Austria, Croatia, Latvia and Argentina.

Paula Fraga-Lamas (Senior Member, IEEE) received her MSc degree in computer engineering from the University of A Coruña (UDC) in 2009, and her MSc and PhD degrees in the joint program Mobile Network Information and Communication Technologies from five Spanish universities: University of the Basque Country, University of Cantabria, University of Zaragoza, University of Oviedo, and University of A Coruña, in 2011 and 2017, respectively. She holds an MBA and postgraduate studies in business innovation management (Jean Monnet Chair in European Industrial Economics, UDC), Corporate Social Responsibility (CSR) and social innovation (Inditex-UDC Chair of Sustainability). Since 2009, she has been with the Group of Electronic

Technology and Communications (GTEC), Department of Computer Engineering (UDC). She has over 70 contributions in indexed international journals, conferences and book chapters, and holds four patents. Her current research interests include Internet of Things (IoT), cyber-physical systems (CPS), augmented/mixed reality (AR/MR), fog and edge computing, blockchain and distributed ledger technology (DLT), cybersecurity, as well as the different technologies involved in mission-critical scenarios under the Industry 4.0 paradigm. She has also been participating in over 30 research projects funded by regional and national government as well as research and development contracts with private companies. She is actively involved in many professional and editorial activities, acting as reviewer, advisory board member, topic/guest editor of top-rank journals and TPC member of international conferences.

Contents

Preface

This book on Computer Security Threats discusses the fundamentals of computer security and presents a broad set of ideas and some of the advanced research in this field. The book is triggered by the understanding that digitalization and growing dependence on the Internet poses an increased risk of computer security threats in the modern world. The chapters discuss different research frontiers in computer security with algorithms and implementation details for use in the real world. Researchers and practitioners in areas such as statistics, pattern recognition, machine learning, artificial intelligence, deep learning, data mining, data analytics and visualization are contributing to the field of computer security.

This book consists of four sections: Section I is an introduction to computer security threats, Section II is on Malwares, Section III is on Botnets, and Section IV is on Blockchain. The book has eight chapters included in these four sections. The chapters include the introduction to computer security threats and techniques that address the threats. Section I is exclusively the introduction to the computer security threats. Section II is on the malicious software or malware, which is the main source of computer security threats. Another major computer security threat is due to botnets and hence Section III is on botnets. Blockchain technology is a decentralized, distributed ledger that records the provenance of a digital asset and its cryptographic algorithm makes it immune to attack. In a world where cyber security has become a key issue, blockchain is a potentially revolutionary technology as it promotes new levels of trust and transparency. Hence, Section IV is specifically on the security aspects of blockchain. Technologies explored in the chapters included in each of these four sections are introduced for the reader in every chapter.

The introductory chapter on computer security threats provides a detailed introduction on the common computer system threats. The logical threats are a main cause of security incidents on computer systems. Knowing these threats and their characteristics helps in identifying the threats and aids in proactively devising steps to protect the systems. The chapter includes motivations and objectives of the hackers. The chapter also includes the classification of threats, which also includes an exhaustive coverage of all the threats. The details of the top security menaces of 2020 are introduced and the expectation for the latter half of 2020 is also discussed.

Malware is the main source of computer security threats and hence the second chapter is specifically focused on the detection of malware embedded into web pages, using client honeypot. In today's world with everyone depending on the Internet, web pages are facing a severe threat of client side browser attacks. The spread of malware uses software vulnerabilities, which attack the client application that sends a request to server. The detection methodology discussed in this chapter is based on client honeypot, which detects the various malicious program linked with web pages. Client honeypots are active security devices in search of malicious servers that attack clients. The client honeypot pretends to be a client and interacts with the server to examine whether an attack happens. Often the focus of client honeypots in on web browsers, but any client that interacts with servers can be part of client honeypot.

Chapter III is on the threats from botnets as various cyber-attacks based on botnets having become one of the most serious security threats on the Internet. Botnet is a common computing platform that can be controlled remotely by attackers by invading several systems called bots in the network space. It is thus an attacking platform consisting of multiple bots controlled by a hacker. As botnets continue to evolve, the behavioral research on botnets is observed to be totally inadequate. The question of how to apply some behavioral problems to botnet research and combine the psychology of the operator to analyze the future trend of botnets is still a challenging issue. As the initial chapter in the section on botnets, Chapter III introduces and discusses the classification and working mechanism of botnets. The chapter also includes the threats evaluation of botnets.

Chapter IV is the evaluation of botnet threats based on evidence chains. At present, although network administrators have firewalls, intrusion detection systems, intrusion prevention systems, and other technical means to achieve partial network protection, they are still confronted with severe challenges in the detection and prevention of botnets, which are known as a threatening attack platform. This chapter proposes a method of botnet threat assessment based on evidence chains. The DS evidence theory is used for network security situational awareness. On the basis of determining the recognition framework, all possible results are considered, and each piece of the evidence is assigned a basic credibility, and the final credibility of the target is fused using the composition rule. The experiments show that this method can work efficiently and detect the major threats in a protected network in real time.

Chapter V is on deploying blockchain in supply chains. In the rapidly evolving environment of the international supply chain, the traditional network of manufacturers and suppliers has grown into a vast ecosystem made of various products that move through multiple parties and require cooperation among stakeholders. Additionally, the demand for improved product visibility and source-to-store traceability has never been higher. However, traditional data sharing procedures in today's supply chain are inefficient, costly, and inadaptable as compared to new and innovative technology. Blockchain technology has shown promising results for improving supply chain networks in recent applications, and has already impacted our society and lifestyle by reshaping many business and industry processes. In an effort to understand the integration of blockchain technology in the supply chain, this chapter systematically summarizes its current status, key characteristics, potential challenges, and pilot applications.

Chapter VI discusses the blockchain applications in cybersecurity. In this chapter the different aspects that relate the application of blockchain with techniques commonly used in the field of cybersecurity are analyzed. Beginning by introducing the use of blockchain technology as a secure infrastructure, the chapter investigates how blockchain can be useful to achieve several security requirements common to most applications. In order to maintain simplicity, this chapter has focused only on some specific cybersecurity disciplines: backup & recovery, threat intelligence, and content delivery networks.

Chapter VII is on the blockchain and the Industry 4.0. It is understood that the extreme automation of factories is necessary in order to face the fourth industrial revolution. This new industrial paradigm will force our industries to manufacture much shorter and customized series at increasingly competitive prices, even tackling the manufacture of thousands of different configurations of a single base product.

In order to achieve this, the production processes must have a flexibility in their configuration that has never been imagined before. This flexibility and ability to adapt automatically to demand is the essence of the fourth industrial revolution and is part of the Western strategy to recover an industrial sector increasingly threatened by the Eastern production of large series at competitive prices. After more than a dozen Proofs of Concept with different manufacturing and energy industries, the chapter describes the scenarios in which blockchain technology brings the greatest benefits to Industry 4.0. After different experiments and through interviews with people in charge of innovation from different industries, the chapter includes an in-depth analysis of the true added value of blockchain in the industry. The outcome is the principal four values of blockchain technology applied to Industry 4.0.

Chapter VIII is on leveraging blockchain for sustainability and open innovation: a cyber-resilient approach towards EU green deal and UN sustainable development goals. In 2015, the United Nations member states identified seventeen Sustainable Development Goals (SDGs) to be fulfilled by 2030. SDGs are an urgent global call for action to provide a blueprint for shared prosperity in a sustainable world. By the end of 2019, the European Green Deal, a roadmap to implement the UN 2030 agenda with a commitment to a growth strategy that will turn environmental challenges into opportunities across all policy areas was presented. To achieve these SDGs, blockchain is one of the key enabling technologies that can help to create sustainable and secure solutions, since it is able to deliver accountability, transparency, traceability, and cyber-resilience, as well as to provide a higher operational efficiency in global partnerships. This chapter overviews the potential of blockchain to face sustainability challenges by describing several relevant applications. The chapter also enumerates different open challenges and recommendations with the aim of guiding all the stakeholders committed to the development of cyber-resilient and high-impact sustainable solutions.

The intended audience of this book will mainly consist of students, researchers, practitioners, data analysts, and business professionals who seek information on the various computer security threats and its defensive measures.

I would like to convey my gratitude to everyone who contributed to this book including the authors of the accepted chapters. My special thanks to the Author Service Manager, Ms. Kristina Kardum and other staff of IntechOpen publishing for their support and efforts in bringing the book to fruitful completion.

Ciza Thomas
Professor,
Directorate of Technical Education,
Government of Kerala,
India

Tiago M. Fernández Caramés and Paula Fraga-Lamas
Group of Electronic Technology and Communications (GTEC),
Faculty of Computer Science,
University of A Coruña,
A Coruña, Spain

Section 1

Introduction

Chapter 1

Introductory Chapter: Computer Security Threats

Ciza Thomas

1. Introduction

Along with the tremendous progress in Internet technology in the last few decades, the sophistication of the exploits and thereby the threats to computer systems have also equally increased. The exploitation is done by malicious hackers who find vulnerabilities or weaknesses, which are the pre-existing errors in the security settings in the computer systems. The common types of vulnerabilities are errors in the design or configuration of network infrastructure, protocols, communication media, operating systems, web-based applications and services, databases, etc.

Threat is a potential risk that exploits a vulnerability to infringe security and cause probable damage/disruption to the information/service stored/offered in/by computer systems or through communication links. A threat to a computer systems occurs when the confidentiality (preventing exposure to unauthorized parties), integrity (not modified without authorization), and availability (readily available on demand by authorized parties) of information on systems are affected. Thus, a computer system threat in general can include anything deliberate, unintended, or caused by natural calamity that effects in data loss/manipulation or physical destruction of hardware. Accordingly, the threats on computer system are classified as physical threats and nonphysical threats. Physical threats cause impairment to hardware or theft to system or hard disk that holds critical data. Nonphysical threats target the data and the software on the computer systems by corrupting the data or by exploiting the errors in the software.

The exploits when successful result in security attacks on computer systems. Hence, threat is a possible danger caused by system vulnerability, while attack is the attempt of unauthorized action or a harmful action. The realization of a threat is usually detrimental and is termed an attack.

In this introductory chapter, the computer security threats are defined as probable attacks from hackers that let them to gain illicit entree to a computer. In this chapter, a detailed introduction is given on the common computer system threats. The logical threats are a main cause of security incidents on computer systems. Knowing these threats and their characteristics helps in identifying the threats and to proactively devise steps in protecting the systems. The organization of this chapter is as follows. Section 2 introduces the motivation and objective of the hackers. Section 3 is on the classification of threats, which also includes an exhaustive coverage of all the threats. The details of the top security menaces of 2020 and the expectation for the latter half of 2020 are introduced in Section 4. Section 5 concludes the chapter.

2. Motivation and objectives of hackers

The purpose of a hacker is to break the security of computers and networks affecting the confidentiality, integrity, and availability of information/service on systems. Such activities of hackers are considered illegal as they invest their time and know how, to make personal gains and breach the security across networks. Before looking at the taxonomy of computer threats, it is necessary to classify the different types of hackers. Each type of hacker is expected to have their own motivation for their activities. The most common of those are included here:

Fun: Fun is the only motivation for the script kiddies and lot of nonserious hackers. For them, the breaking into a secure system is a challenging and adventurous enjoyable game to test their wits and skills.

Vulnerability testing: Vulnerability testing is done by administrators to locate vulnerabilities and hence develop protections. The same is also done by hackers to identify vulnerabilities in target systems and to find the exploits for those vulnerabilities. This is almost a pre-phase of an attack.

Theft: Theft or stealing of data is when hackers infiltrate on a database of credentials of individuals or organizations.

Espionage: Espionage is another type of theft where the hacker tries to get protected information instead of the direct financial gain. The information stolen can be either sold in black market or used by adversaries to gain strategic advantages.

Spamming: Spamming is not just about unsolicited emails. This spam can be due to certain particular malware that invade the web browser and devastate with unwanted ads.

Control: The hacker uses a Trojan or other means to take remote control over another system. Then the hacker can turn that compromised system into a bot or a zombie computer that they use to power spam or to deploy distributed denial of service attacks.

Disruption: Disruption of services or access to information, by taking over websites or social media accounts, is usually an act of competition, protest, or rivalry. This effect will slow down or shut down of the target's Internet activity.

3. Classification of computer threats and attacks

Computer threats and attacks involve accessing information, obliterating or manipulating data, destabilizing the computer, or degrading its performance [1]. Computer attacks are mainly information gathering, privilege escalation, buffer overflow exploits, remote accessing by unauthorized users, and denial of service attacks [2]. Network attacks being a subset of computer attacks were mostly attacks on computer systems that form the basic infrastructure of a communication network. A network aids in sending an attack or it could be the means of attack.

There are various steps involved in the attacking scenario, and these steps are briefly listed here:

Step 1: spoofing

Before initiating any of the attacking steps, the hackers normally prefer to hide their identity and their activities. These are normally done by spoofing when the attacker hides his identity and pretends to be someone else. This can be done by MAC cloning, IP spoofing, or email spoofing.

Step 2: reconnaissance

It is always a good practice to plan well before undertaking any action, and this is applicable in the case of hacking too. The hackers first identifies a target to launch an attack, extract maximum information regarding this target, understand its vulnerabilities, and then only explore the best ways to exploit it.

Step 3: weaponization

The hacker with the information collected in the previous phase identifies/ develops weapons in order to get into the computer or the network. During this phase, the hacker collects the tools that they plan to use once they gain access to the system for the successful exploitation of the vulnerabilities in the system.

Step 4: implementation

In the implementation phase, the attack starts working. It is when the phishing e-mails are sent or when the fake web pages are posted to the Internet and the attacker patiently waits for all the data they need to start rolling in.

Step 5: exploitation

This is a state when the sensitive and confidential data starts rolling in. It is the most exciting phase for the hackers, and they try out the usernames and passwords against web-based e-mail systems or secured connections to sensitive networks.

Step 6: installation

After a successful exploitation, the attacker will make sure to have continued access to the system. This is by installing a persistent backdoor or creating admin accounts on the system, disabling firewall rules, and perhaps even activating remote desktop access on computer systems on the network.

Step 7: control

Once the attacker gains access to the network or creates administrator accounts or installs all the necessary tools for backdoor entry any time to the system, the attacker is in control of the target.

Step 8: action on set goals

With total control on the target system, the attacker can set goals and achieve it with or without the knowledge of the genuine user.

The attacks are thus classified depending on the various steps taken by the hacker in the process of the attack, starting from hiding the identity to information collection, which is the pre-phase of an attack, to the actual attack.

4. Computer threats

4.1 Spoofing

Spoofing is when someone hides their identity to evade detection for their wrong acts and pretends to be someone else in an attempt to gain trust and get sensitive system information. The common spoofing done by changing the hardware or MAC address is called MAC cloning, changing the IP address or the unique identity on the network is called IP spoofing, and impersonating as someone else in their digital communication is called email spoofing.

4.2 Information-gathering attacks

Information gathering is the practice of attacker gaining priceless details about probable targets. This is not an attack but only a pre-phase of an attack and is totally passive as there is no explicit attack. Systems including computers, servers, and network infrastructure, including communication links and inter networking devices, are sniffed, scanned, and probed for information like whether the target system is up and running, what all ports are open, details regarding the operating system and its version, etc. Some of the information-gathering attacks are sniffing, mapping, vulnerability scanning, phishing, etc.

4.3 Password attacks

The simplest way to achieve control of a system, or any user account, is through a password attack. If the personal and behavioral details of the victim are known, the attacker starts with guessing password. Frequently, the attacker uses some form of social engineering to trace and find the password. Dictionary attack is the next step in password attacks and is automated.

4.4 Malware

After gaining access to a system, the attacker takes the support of malware or malicious software that clandestinely acts against the interests of the computer user.

4.5 Virus

Computer viruses are the most communal threat to the computer users. Computer viruses are malicious software designed to blow out from one computer to another through file transfer, piggybacks on genuine programs and OS, or e-mails. The email attachments or downloads from particular websites contaminate the computer and also other computers on its list of contacts by using the communication network. Viruses influence the system security by changing the settings, accessing confidential data, displaying unwanted advertisements, sending spam to contacts, and taking control of the web browser [2]. The viruses are identified as executable viruses, boot sector viruses, or e-mail viruses.

4.6 Worms

Computer worms are fragments of malicious software that reproduce swiftly and blow out from one computer to another through its contacts, again spreading to the contacts of these other computers and so on and reaching out to a large number of systems in no time. Captivatingly, worms are prepared for spreading by exploiting software vulnerabilities. Worms display unwanted advertisements. It uses up tremendous CPU time and network bandwidth in this process thereby denying access to the systems or network of the victim, creating chaos and trust issues on a communication network.

4.7 Trojans

Trojans are programs that appear as perfectly genuine but, in reality, have a malicious part embedded in it. Trojans are spread usually through email attachment from the trustworthy contacts and also on clicking on fake advertisements. The payload of Trojans is an executable file that will install a server program on the victim's system by opening a port and always listening to that port whereas the server is run on the attacker's system. Hence, whenever the attacker wants to login to the victim machine, they can do so by means of the backdoor entry making it hidden from the user.

4.8 Spyware and adware

Spyware and adware are software with a common property of collecting personal information of users without their knowledge. Adware is intended to track data of the user's surfing behaviors, and, based on that, pop-ups and advertisements are displayed. The adware clause in the agreement during the installation process is often skipped with least seriousness. Spyware on the other hand gets installed on a computer and gathers information about the user's online activities without their knowledge. Spyware contains keyloggers that record everything typed on the keyboard, making it unsafe due to the high threat of identity mugging.

4.9 Scareware

Scareware is yet another malware that tricks victims by displaying fake alerts and forcing the victim to buy protective software that is fraudulent. The alerts or the pop-up messages sound like warning messages along with proper protective measures, which if followed creates security issues.

4.10 Rootkit

Rootkit is a pool of software tools that gets mounted in stealth along with some genuine software. Rootkit allows remote access and administrative control on a system. With these privileges, the rootkit performs malicious activities like disabling of antivirus, password sniffing, keylogging, etc.

4.11 Keylogger

Keylogger software has the ability to record keystrokes and also capture screenshots and save it to a log file in encrypted form. Keylogger software can record all the information that is typed on the keyboard including passwords, e-mail, and instant messages. The log file created by the keylogger is saved and mailed to the attacker on a remote machine with the motive to extract password and banking details for financial fraud.

4.12 Ransomware

Ransomware is a malicious software that hampers admission to computer or files on the computer. The computers may be locked or files encrypted. Accordingly, the two common types of ransomware are lock screen ransomware and encryption ransomware. The victim will be demanded ransom for the restriction to be removed, and this gets displayed on victim's system. There can also be notification stating that establishments have detected illicit activity on this computer and demands ransom as fine to avoid prosecution.

4.13 Rogue security software

Rogue security software is another malicious program that deceives users to believe that there is malware installed on their system or the security measures are outdated and hence of concern. They offer installing or updating users' security settings. Then it is an actual malware that gets installed on the computer.

4.14 Botnets

A collection of compromised systems or bots acts as a team of infected computers under the control of a bot master to remotely control and send synchronized attacks on a victim host. This army of bots, agents, and bot master constitute a botnet. Botnets are used for sending spams and also for distributed denial of service attacks.

4.15 Denial-of-service attacks

Denial-of-service (DoS) attacks as the name suggests deny users from accessing or using the service or system. This is mainly done by overwhelming the bandwidth, CPU, or memory wherein the access to the network of the victim machine or server offering the service gets denied. DoS attacks thus interrupt the service of a computer or network systems, making it inaccessible or too inferior in performance.

4.16 Distributed DoS

In distributed DoS (DDoS) attacks, the victim is targeted from a large number of individual compromised systems simultaneously. The DDoS attacks are normally done with the help of botnets. The botmaster is the attacker who indirectly attacks the victim machine using the army of bots or zombies. The DDoS attacks occur when a large number of compromised systems act synchronously and are being coordinated under the control of an attacker in order to totally exhaust its resources and force it to deny service to its genuine users. It is the upsurge in the traffic volume that loads the website or server causing it to appear sluggish [2].

4.17 IoT-based attacks

The last decade has seen exponential increase in the use of Internet of Things (IoT) that are smart devices used at home, organizations, and businesses. The issue with these IoT is its weak security as these devices are often overlooked when it comes to applying security patches that create lead-ins for attackers to seize these devices to infiltrate the networks. An IoT-based attack is any cyberattack that leverages a victim's use of IoT to sneak malware onto a network.

4.18 Session hijacking

In session hijacking, the hacker takes control of a session going on between two hosts. Session hijacking usually takes place in applications that use TCP with a sequence number prediction. With that sequence number, the attacker sends a TCP packet.

4.19 Blended attacks

A blended attack is a software exploit that encompasses a mixture of exploit techniques to attack and propagate threats, for example, viruses, worms, and Trojan horses.

4.20 Website attacks

Website attacks are targeting browser components that are at risk of being unpatched even when the browser is patched. SQL injection attacks are intended to target any website or web application that uses an SQL database such as MySQL, Oracle, etc. by taking advantage of the security flaws in the application's software. This attack is used to obtain and corrupt user's sensitive data.

4.21 Mobile phone and VOIP threats

Malware target mobile phones, VoIP systems, and the IP PBXs as these devices have plentiful published vulnerabilities. There are attack tools freely available on the Internet, and misusing these vulnerabilities makes these attacks too common and simple even for a script kiddie.

4.22 Wi-Fi eavesdropping

Wi-Fi eavesdropping is an attack used by network attackers to grab sensitive information of a target system. It is the act of silently listening on an unencrypted Wi-Fi network.

4.23 WPA2 handshake vulnerabilities

The key reinstallation attack (KRACK) lets an attacker to decipher the network traffic on Wi-Fi routers. Every device connected to Wi-Fi, such as computers, smartphones, smart devices, and wearables, can be identified by the hacker.

4.24 Insider attacks

One of the prevalent all-time computer security threats faced by any organization is from its own employees. Insider attacks are initiated by disgruntled employees of an organization. Insider usually has certain privileges to the data as well as rights on the systems and networks that they attack, giving them an advantage over external attackers. These attacks can be hard to prevent with firewalls, which are the first level of defense.

4.25 Supply chain attacks

A supply chain attack seeks to cause harm by targeting the least secured elements in the supply network.

4.26 Buffer overflows

Buffer overflows are used to exploit programming glitches that do not take care of the buffer size. If a buffer is jam-packed beyond its size, the data overflows into the contiguous memory. This flaw gets smartly used by hackers to change the execution of the program.

4.27 User to root attack

User to root attack is a case of privilege escalation where a user gains a higher privilege than that authorized. This is not a class of attack as such, and it is the process of any attack. Every attack will do activities the attacker is not privileged to do.

4.28 Man-in-the-middle attacks

Man-in-the-middle attacks allow the hacker to snoop on the communication between two systems, affecting the privacy. A common method of doing this is to place the attacker at a point and redirect all the communication through the route that includes that hacker so that eavesdropping is possible by the hacker.

4.29 Pharming

Pharming is a widespread online fraud that will automatically point to a nasty and illicit website by relaying the authentic URL. Even when the URL is correctly entered, the redirection happens to some forged website looking similar to the actual one. This fake site prompts one to enter personal information that gets to someone with a wicked intent.

4.30 Spam

Spams are unsolicited bulk e-mail messages that annoy the user with unwanted and junk mails. It gives burden for communications service providers, organizations and individuals alike. These emails can be commercial ones like an advertisement or noncommercial one like chain letters or anecdotes. Spam is considered an active vehicle for virus propagation, scams, fraud and is a threat to computer privacy. Spam also phishes for interesting information with offers and promotions that trick victims into following links or entering details.

5. Present-day computer security threats and trends

Predicting the computer security threats and trends is usually done to lend a hand to the security experts who take proactive measures to protect security. Normally the predictions for any year depends on how it went in the previous years, and the changes expected are mainly in terms of the tactics and scale of the biggest and significant threats that were successful in implementation and also in evading detection. The investment on security is justified in many organizations only after analyzing these predictions.

Phishing and other social engineering tactics are likely to continue in the coming years too with increased complexity and sophistication. They will appear to be more and more convincing to trick people into clicking on a link or opening attachments. Even with strong defenses to protect against ransomware, hackers are expected to all the time target more victims with large digital assets. The rise of cryptocurrency like bitcoin will also trigger more ransomware attacks by letting demands for payment made incognito. Cryptojacking can also be seen as a common trend of future as it involves hackers hijacking with a purpose of mining for cryptocurrency.

As the Internet of Things is becoming widely popular and more ubiquitous, the IoT attacks will be on the upsurge. IoT includes laptops, tablets, smart wearable devices, webcams, household appliances, Wi-Fi-enabled speakers, appliances, alarm clocks, medical devices, manufacturing equipment, automobiles and networking devices like routers, gateways, switches, NAS servers, and even home security systems. Security is rarely the first concern in the competition to bring new products and technologies. Thus the more IoT devices, the greater the risk, making IoT attacks to be on the rise in coming years.

Data breaches will continue in the coming years as data remains a valuable black market attraction.

Totally new approaches for data and infrastructure protection are essential as more and more data is moved to the cloud. Also, in the coming years, there will be more attacks targeting electrical grids, automated transportation systems, computerized water treatment facilities, etc.

State-sponsored attacks are when states or nations are using their cyber skills to infiltrate other governments and execute attacks on severe infrastructure. As political strains grow, state-sponsored attacks steal political and industrial secrets, spread misinformation, perform DDoS attacks, execute prominent data breaches, etc.

Another target of attacker is the all-time sensitive medical record of patients. As the healthcare industry gets used to the digital age, concerns around privacy, safety, and computer security threats are also seen to rise. There are worries about a hacker taking over and changing dosages of medicines, disabling vital sign monitoring, etc., as these are life-threatening to the patients.

Now, with the self-driving cars, semiautonomous vehicles, and the connected cars, the risk of cyber security is stringent and serious. With high-tech automobiles, the future will likely see an increase in not only the number of connected cars but in the number and severity of system vulnerabilities detected. For hackers, this means yet another opportunity to exploit vulnerabilities and cause threat to life.

Endpoint security will be a major concern for organizations as malware infections of employee-owned devices are going to be a major security issue in 2020 when employees start "working from home" in the wake of COVID 19 pandemic. When organizations permit employees not to risk their health and safety and allow them to use their own devices, attackers will target those devices to bypass the multilayered defenses of the organization. The advantage to hackers is that the users' personal devices are less protected compared to corporate devices as users rarely apply added measures to protect their smart devices from impending threats.

Artificial intelligence also gets applied on both sides of the barricade for protecting and attacking the computers. Artificial intelligence is being used for person identification, threat detection, etc. to aid security; however it is also being weaponized by hackers to develop increasingly complex malware and attack methods.

6. Conclusion

A lot of computer threats have been included in this chapter with many terms tending not to be mutually exclusive. Again, an attack may get classified into different classes since attackers use multiple techniques or strategies. The irony is that even with lot of advanced defensive mechanism put in place by security experts, the hackers may still use the same attacking techniques and will take advantage of the same vulnerabilities they have used in the past. It is important to defend the attacks by paying attention to the internal systems, deploying multiple defenses for enhanced security, and avoiding irreparable damage. This requires the implementation of security policy as an ongoing process with tight access control mechanism and deployment of advanced multiple layer security devices.

Author details

Ciza Thomas
Directorate of Technical Education, Government of Kerala, India

*Address all correspondence to: cizathomas@gmail.com

References

[1] Thomas C, Balakrishnan N. Improvement in intrusion detection with advances in sensor fusion. IEEE Transactions on Information Forensics and Security. 2009;**4**(3):542-551

[2] Thomas C. Performance enhancement of intrusion detection systems using advances in sensor fusion [Ph. D dissertation report]; 2009

Section 2

Malware

Chapter 2

A Detection of Malware Embedded into Web Pages Using Client Honeypot

M. Veena, S. Upasana, S. Prathima and Sudha Senthilkumar

Abstract

In today's Internet world, web pages are facing a severe threat which uses the client-side browser attacks. The vulnerability-based attacks are based on client-side application which becomes the major threat to web pages. The spread of malware uses software vulnerabilities which attack the client application sending request to the server if whether the attack has occurred. This detection is based on client honeypot which detects the various malicious program linked with web pages. Client honeypots are active security devices in search of malicious servers that attack clients. The client honeypot poses as a client and interacts with the server to examine whether an attack has occurred. Often the focus of client honeypots is on web browsers, but any client that interacts with servers can be part of client honeypot. In this research paper, we propose a model of detecting embedded web pages using client honeypot.

Keywords: client honey pots, network, malware, cyber space, client-side attack

1. Introduction

The Internet has become the most popular medium where the increasing new trend has availability of spreading attacks. There are existing attacks as firewalls and intrusion detection system, and honeypot is also one of the technology-based security attacks. Here honeypots are playing a big role in security attacks, and it is a new kind of attacks on the cyber space. There are many attacks in the area of security. Honeypots refer to closely monitoring system which needs to be attacked. To supplement a new value it must be compromised, attacked by cyber crimes on honeypot.

Honeypots are based on servers which will not be able to detect the client-side applications in web pages. Honeypots crawl the network of any firewall that attacks the client. There are two attacks: client and server. Server honeypots is a traditional honeypot, whereas client honeypot is based on client-side scripting on web pages. This attack detects the attack from the client side which is vulnerable on the client side. It needs a source and visits and detects all activities. It spreads malware through the vulnerability in the client-side attack. Here we are using only client honeypot and vulnerability-based attack. The vulnerability is based on exploiting the attack where the security detects the services exposing on attacking the system.

Client honeypots crawl the network, interact with servers, and classify servers with malicious web pages. It does not expose server-based attacks on the client-side

IntechOpen

attack. They have some kind of exposed attack which is vulnerable to the sender and receiver. They can be detected, if they are passive.

2. Proposed work

The security resources are production value; no resources should communicate between each other. Honeypot is compromised for outbound connections on the web pages. Honeypot collects all information about the intruder or intermediate where the community is targeting to attack. And they list the type of resources attack on the network security. Honeypots play the big role on the attacker side scripting based on web pages in client-side attack. Client honeypots are also called as active honeypots or honey client. It visits the web page as requested by the attacker and visits the web page to check whether the attack has happened or not [1–3] (**Figure 1**).

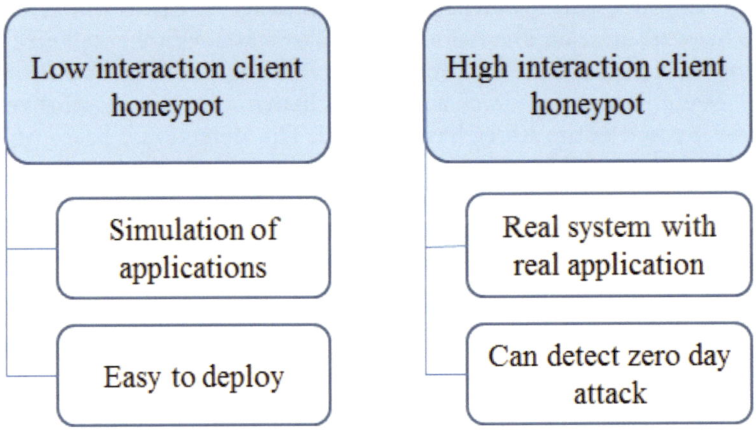

Figure 1.
Client honeypot classifications.

3. Client honeypots

A honeypot is one of the security technologies that helps an organization to catch viruses, malware, or attackers, and it acts as an alarm system that discovers the attempts to attack a network. Honeypot technology is defined as a "security resource whose value lies in being investigated, attacked, or compromised" [1]. The types of honeypot are active and passive. A technology that passively waits for attacks to detect them are called passive honeypot. Active honeypot also called as client honeypot that interacts with a target web page to find its possible effect on the honeyware.

4. Architecture

The client honeypot architecture is separated into three components, namely, queuer, the client, and analysis engine. A queuer is a process of creating a list of servers for the client to visit. A client who can able to create request to servers is recognized by the queuer. An analysis engine is a process of identifying an attack processing in client honeypot. Along with all the above components, client honeypot

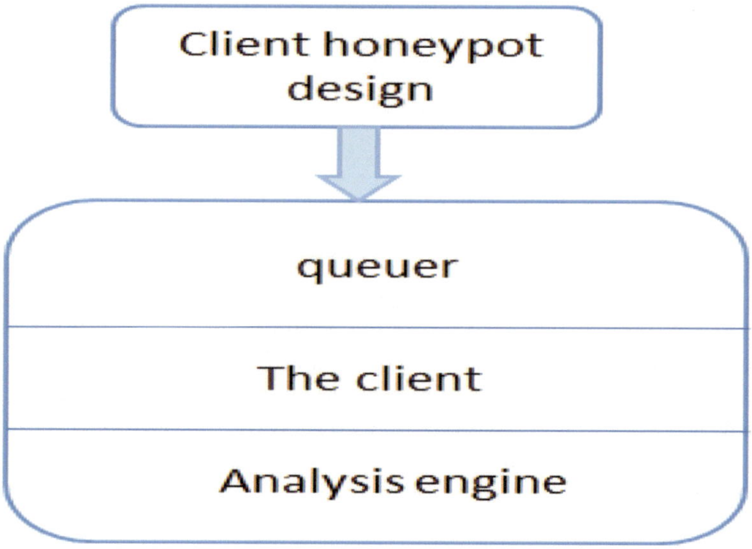

Figure 2.
Client honeypot design.

is furnished with some kind of approach to avoid successful attacks from exploring beyond the client honeypot [4, 5]. Analog to traditional server and client honeypots are classified by their high- or low-interaction level that denotes the client honeypot make utilize of functional interaction the server. This is a newly hybrid approach that uses both high- and low-interaction detection techniques (**Figure 2**).

5. Client honeypot solutions

5.1 High-interaction client honeypots

High-interaction client honeypot is a real application installed on the real systems. Real browsers and plug-ins are being browsed by the websites. Attacks are detected by checking the state of the process after a server interaction. Capture differentiates from existing client honeypots in different ways. It is designed to be fast and to be scalable. Event-based model allow to know the detection of state changes. A main capture server can able to manage several clients across the network.

Honeyclient is a web browser. It is an open-source honeypot and a mix of perl, c++. It detects attacks on Windows client by registry entries, monitoring files, and processed events. It included the capture-HPC. It also contains a crawler, so that it can be sowed with a list of URLs from start and continues to exchange web pages in search of client-side malware. HoneyMonkey is also a web browser. It is not an open source. It detects attacks on Windows client by registry entries, monitoring files, and processed events. It is a layered approach to communicate with servers to identify zero-day exploits. If the attack is still identified, one can complete the attack as no patch has been publicly released and it is dangerous [6–8]. SHELIA is a combination of the process of email received and email reader. It opens different client applications depending on the type of URL or the received attachment. It observes the executable instructions that are processing in data area of memory that indicates a buffer. UW Spycrawler is integrated; with the web browser like Mozilla, it cannot be downloaded. It detects attacks on Windows client by registry entries, monitoring files, browser crashes, and processed events. Event-based mechanism is used to detect by spcrawlers [9, 10]. It increases the time period of the virtual machine. It is a process to overcome time bombs. WEF is an automatic implementation of drive by download that detects in virtualized environment. WEF is used as an active HoneyNet with overall simulated architecture beneath for rollbacks of compromised virtual machines.

5.2 Low-interaction honeyclient

Low-interaction honeyclient is different from high-interaction honeyclient in that they do not use the entire real system. But it uses lightweight or simulated clients to communicate with the server. Responses received from servers are scanned directly to consider whether an attack has been taking place or not. It is a platform-independent open-source framework written in Ruby [11]. It concentrates on driving a web browser emulator which interacts with the server. Mischievous server is identified by statically investigative the web server's response for mischievous string through the usage of snort signatures. Honeyclient uses many existing freely available open-source software systems. It consists of the following components:

5.2.1 Queue/seed generation

It is a source of initial set of seed URLs.

5.2.2 Web search seeding

The three web search engine application interface are Google, Yahoo, and MSN which are common keywords.

5.2.3 Spam trap seeding

Spam mails are extracted from URLs and are enlisted.

5.2.4 Blacklist seeding

It is a tool designed to automatically download blacklist from major blacklist workers and seed for crawler [12, 13].

5.2.5 Web crawling

Heritrix crawler is simulated into the monkey spider prototype with two parameters predefined:

- Maximum link hops which counts the connections to be included in crawl
- Maximum transitive hops which count the URLs extracted from seeded URLs

5.2.6 Content/malware analysis

5.2.6.1 Static analysis

The contents that are downloaded from the URL are scanned by ClamAV antivirus and it alerts using pattern matching. The terminology is provided for the downloaded binary [14, 15].

5.2.6.2 Dynamic analysis

Malware analysis tool like CWSandbox is performed [16].

It is implemented to copycat the behavior of a user-driven network client application and abused by an attacker's content. It is a virtual honeyclient which means that it is not a real application but it is an emulated client. It performs dynamic analysis of JavaScript and visual basic scripts to delete the complication from malicious pages. To analyze the malicious content, complication or encrypted JS is decrypted and reanalyzed. SPYBYE allows a web master to identify whether a website is eroded by a set of heuristics and scanning of data against the clamAV. It is a tool that communicates with a URL that is integrated with a web browser through its user agent field and downloading the response of the target website. The response is exploited using the scan engine [17, 18].

5.3 Problem statement

The problem explained in this paper is to identifying and extracting malicious web programs combined into malicious websites on the client honeypot technology. It is used to address the problem to determine malicious websites based on the maliciousness inside the websites. It helps to rectify the difficulty of identifying whether we need to visit the website or not. Client honeypot technology can able to identify the client-side attacks by surfing the web pages [19, 20].

During the implementation, the steps of generic algorithm are used in the URL separation and feed within the virtual machine for visit:

1. List of URL to be visited, say N numericals of URLs.

2. Store the list of URLs into the database.

3. Get the URL one by one from mysql database.

Figure 3 represents the process of the system listing websites which have been presented as having visited a true browser on a client honeypots. By using clean machine of client honeypots, after visiting the clean website taking snapshot of that machine and store the log created at the stage of website visit.

System logs are created in client machine that are analyzed using third-party analyzing tool like antiviruses to identify the contagions on the list of collected logs.

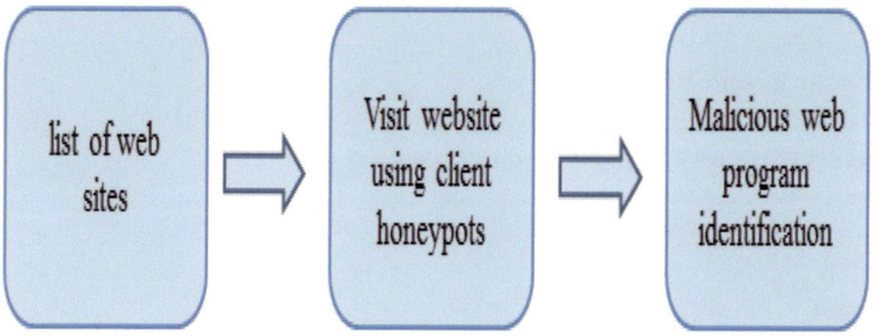

Figure 3.
Process of identifying malicious program.

6. Conclusion

In this paper, we presented a study of current solution of client honeypots for identifying the mischievous websites, and it is not applicable for closely bound and public users. So we propose a system which is able to identify the malware platforms with the help of client honeypot and put on the clever forensic inquiry of the collected network data.

Author details

M. Veena[1], S. Upasana[1], S. Prathima[1] and Sudha Senthilkumar[2*]

1 VIT University, Vellore, India

2 School of Information Technology and Engineering, VIT University, Vellore, India

*Address all correspondence to: sudha.s@vit.ac.in

References

[1] Sun X, Wang Y, Ren J, Zhu Y, Liu S. Collecting Internet malware based on client-side honeypot. In: 9th IEEE International Conference for Young Computer Scientists (ICVCS 2008); 2008. pp. 1493-1498

[2] Spitzner L. Honeypots: Tracking Hackers. Addison Wesley; 2002

[3] Honeynet Project. Know Your Enemy: Defining Virtual Honey/nets. 2003. Available from: http://old.honeynet.org/papers/virtual/

[4] Danford R. —2nd Generation Honeyclients. SANS Internet Storm Center; 2006. Available from: http://handlers.dshield.org/rdanford/pub/Honeyclients_Danford_SANS 06.pdf

[5] Available from: http://www.honeyclient.org/trac

[6] Available from: https://projects.honeynet.org/capture-hpc

[7] Available from: http://en.wikipedia.org/wiki/HoneyMonkey

[8] Available from: https://projects.honeynet.org/honeyc

[9] Available from: http://www.xnos.org/security/overview

[10] Available from: http://code.google.com/p/phoneyc/

[11] Available from: https://en.wikipedia.org/wiki/Client_honeypot

[12] Available from: http://www.spybye.org/

[13] Alosefer Y, Rana O. Honeyware: A web-based low interaction client honeypot. In: Third International Conference on Software Testing, Verification, and Validation Workshops (ICSTW); 2010

[14] Available from: http://www.honeynet.org/node/158

[15] Ramaswamy C, Sandhu R. Role-based access control features in commercial database management systems. Citeseer 1998

[16] Skoudis E, Zeltser L. Malware: Fighting Malicious Code. Prentice Hall; 2003. p. 3

[17] Available from: http://monkeyspider.sourceforge.net/

[18] Available from: http://www.cs.vu.nl/~herbertb/misc/shelia/

[19] Provos N, McNamee D, Mavrommatis P, Wang K, Modadugu N. The ghost in the browser analysis of web-based malware [Online]. 2007. Available from: http://www.usenix.org/events/hotbots07/tech/full_papers/provos/provos.pdf [Accessed: 11 February 2009]

[20] Secure Browsing. Malware Protection. Trustwave. Available from: https://www.trustwave.com/securebrowsing/

Section 3

Botnets

Chapter 3

Threats from Botnets

Ji Yimu and Liu Shangdong

Abstract

At present, various cyberattacks based on Botnet are the most serious security threats to the Internet. As Botnet continue to evolve and behavioral research on Botnet is inadequate, the question of how to apply some behavioral problems to Botnet research and combine the psychology of the operator to analyze the future trend of Botnet is still a continuous and challenging issue. Botnet is a common computing platform that can be controlled remotely by attackers by invading several noncooperative user terminals in the network space. It is an attacking platform consisting of multiple Bots controlled by a hacker. The classification of Botnet and the working mechanism of Botnet are introduced in this chapter. The threats and the threat evaluation of Botnet are summarized.

Keywords: Botnet, threat evaluation, Botnet classification, Botnet threat

1. Introduction

In 1990, the continuous development of the global economy led to the continuous reform and innovation of information technology, which gave birth to the computer and the Internet, and Internet technology was introduced into every household. In the new century, with the globalization and informatization of network, computer network has become indispensable knowledge for the development of the Internet. At present, the demand of computer network is increasing, and various social organizations such as enterprises, governments, and schools are constantly connecting themselves to the network to exchange and share information resources. With the interconnection of global networks, the Internet is everywhere in the world. From China's core report the 43rd Statistical Report on Internet Development in China [1], we can see the development of the Internet in China and the country's emphasis on the Internet.

The number of Internet users in China has increased gradually from 2007 to 2019, reaching more than 829 million in 2018. The penetration rate of the Internet also increased dramatically year by year. In 2018, the penetration rate was more than 59.6% for the population. It can be seen that the resources of the Internet are accessible to everyone.

The emergence of the Internet has brought a lot of convenience to people's life, but meanwhile, with the continuous expansion of network scale, the security risks have been exposed. For the computer network itself, there are some inherent security risks in design. With the network scale gradually expanding and complex network environment, many criminals make use of the vulnerability on the network for network invasion, information leakage, hacker blackmail, and other attacks. These hazards not only affect people's safe use of the network but also can lead to the disclosure and destruction of sensitive information of enterprises, public institutions, military, and financial institutions, adversely affecting the national economy and security. According to the data of the 2018 China Internet Cyber Security Report provided by the China National Internet Emergency Center [1], the number of security vulnerabilities collected by the National Information Security Vulnerability Sharing Platform is 14,201 in 2018.

Botnet is a common computing platform which can be controlled remotely by attackers by invading several noncooperative user terminals in network space. "Invading in network space" refers to an area where hackers can enter and exit at will to send arbitrary information and files within an IP block or an Internet region; "noncooperative" means that a vulnerable computer receives no warning notice for the upcoming attack; and "remote control" means that a Botnet usually has a C&C server that can remotely accept control commands from hacker and concurrently send the corresponding instructions in the form of messages to the corresponding infected host (Bot). Over time, a small Bot can be expanded to be a Botnet with thousands of Bots, which, due to the large number of Bots, has high-performance storage size and fast computational response time. Making use of these characteristics, hackers can easily occupy network flow and launch corresponding persistent attacks on a specific target, such as mail attacks, HTTP flooding attacks, etc. At this stage, Botnet has become the main attacking method used by hackers. Due to its simple formation and various types, Botnet has become one of the biggest threats to Internet security and a key research topic by experts.

A Botnet is an attacking platform composed of multiple Bots that is controlled by the commands that hackers send to it, and its behavior is also controlled by hackers. Therefore, the attack of Botnet is generally controlled by the subjective consciousness of the hacker, which leads to the threat generated by it making it hard to locate and predict its threat. From the last century to the present, Botnet attacks not only cause network equipment paralysis but also seriously affect the country at political and economic level, involving military aspects as well. Many newspapers and magazines have published Botnet attacks. In the early twenty-first century, the Conficker Botnet, which was spread by network sharing and U disk, has spread tens of thousands of host computers, and this Botnet mainly made use of the vulnerability MS08–067. During that attack, not only the personal computer was affected, but also the national defense platforms of Germany and the United Kingdom were affected to varying degrees. Some aircrafts were delayed because the attacks prevented releasing of normal commands. In 2016, the United States experienced a large area of network outage, which was caused by a denial-of-service attack on Dyn, a famous American company. The company emphasized that the attack covers millions of IoT devices around the world (the source IPs of UDP/domain name server (DNS) attack are almost fake IPs, so this number does not represent the number of Bots) and some of the important attacks are from IOT devices. Through analysis, the culprit of the incident was the Mirai Botnet, whose source code was published online [2]. According to the 2017 China Internet Security Report, more than 200,000 IP addresses in the Chinese mainland have been affected by hacker attacks, including more than 4000 C&C servers serving to convey commands. These cases show that Botnet poses a serious security threat to China.

China, even the whole world, has paid great attention to the security problems caused by Botnet. In the field of scientific research, on January 23, 2008, the "Seminar of Response to Botnet" sponsored by China National Internet Emergency Center/Coordination Center (CNCERT/CC) was held in Huaxin Building, Beijing. At the International Supply Media Conference held in Nice, France, in 2017, Derek Manky, head of global security strategy of Fortinet, said that the intelligent cluster networks could replace Botnet as a new threat in the future. At the 8th International Conference on Communication and Network Security (ICCNS) in 2018, research topics such as communication and network security, malware and Botnet, and communication privacy and anonymity were discussed in depth.

There are several reasons why Botnet can become the biggest security threat in the world:

1. The development history of Botnet is divided into two phases. It mainly is a kind of virus or worm in the first phase and transforms into the Botnet platform in the second phase. The advantages of the virus are rapid infection and rapid transmission, but the disadvantages are also obvious, that is, the Bot cannot be controlled by the hacker, the degree of infection cannot be perceived by the hacker, and the infected geographical area is very limited and cannot be expanded on a large scale. In summary, the virus is small scale but uncontrollable. The Botnet combines the advantages of the virus and overcomes the shortcomings of virus, so it is very popular among hackers.

2. The virus attack has the characteristic of integration. Botnet is different, the control command of Botnet is issued by separate C&C server, and the attack and invasion are completed by the controlled Bot. The C&C server and the controlled host will make requests and connections through HTTP packets. In this way, hackers only need to send a few commands to the C&C server to launch diversified forms of attack, which improves the flexibility of Botnet and enhances the concealment of Botnet.

3. Security is the foundation of each computer field, and the development of any field will be accompanied by technical achievements in the security of this field. Because Botnet and security measures are developed in a certain order, Botnet can rise rapidly during this period. In the expansion process of Botnet, the first thing is to find the C&C server, and the hackers will make use of the vulnerability to snatch the control of the host. For example, Mirai Botnet will use the weak password vulnerability to hack into the server's telnet port to gain control of the host; the IRC Botnet will break the shared chat room server for the construction of its own C&C server; due to lack of security awareness of users, some companies' cloud servers are also hacked by hackers and used as C&C server, such as Alibaba Cloud, Tencent Cloud, etc.

4. The Botnet applies the knowledge of the key to the management of the Botnet controller in order to prevent the entire Botnet from being uncontrollable after the C&C server is compromised by security experts, so as to improve its concealment and survivability. For example, in a decentralized Botnet, multiple C&C servers are used for unified control, and encryption technology and authentication technology are used in the process of message transmission between C&C servers; in this way, illegal messages cannot be accepted by the controller so as to prevent replay attacks.

Through the above analysis, the process of defending Botnet can be summarized into five steps: analysis and detection, trusted tracking, measurement, situation prediction, and counterattack. Among them, the "analysis and detection" is to find cues of Botnet from the data flow; the "trusted tracking" is to determine the information source of the Botnet; the "measurement" is to manipulate the architecture, life cycle, and attack process of the Botnet; the "situation prediction" is to evaluate the next activity of the Botnet in advance and to prevent and warn in advance; and the "counterattack" is to reduce its activity and break the C&C server to paralyze the Botnet.

At present, there are many different methods for detecting Botnet. For example, Moheeb and others built a real network flow monitoring system to analyze the flow records, binary file types, Botnet control commands, etc.; Cai [3] evaluated the key behavioral characteristics of HTTP Botnet and designed a detection method for HTTP Botnet based on feature analysis; Song [4] adopted displacement entropy and Kalman filtering to detect and analyze the characteristics of P2P Botnet and proposed the corresponding detection algorithm; XU found that P2P Botnet shows higher robustness when random nodes fail, but the robustness declines rapidly when central nodes fail; and Chen proposed a solution to the problem that HMM method cannot be adopted for flow detection of hierarchical Botnet.

2. Classification of Botnet

Botnet has many types of classification, and it can be divided into centralized Botnet and distributed Botnet according to different operating principles.

1. For centralized Botnet, there is only one C&C server in the whole Botnet platform, and all Bots are connected to the C&C server. C&C server has the right to control all Bots.

2. For distributed Botnet, the Bots will also have message communication between each other. According to different command and control protocols, the centralized Botnet can be classified into three categories: IRC-based Botnet, HTTP Botnet, and custom protocol Botnet [5–7]. According to topological structure, the distributed Botnet can be classified into three categories: structured P2P Botnet, unstructured P2P Botnet, and hierarchical Botnet [8, 9]. **Table 1** lists the classification of some known Botnets. Although there are multiple control servers in some Botnets, such as Mega D and Mariposa [10], Bots do not communicate with each other, and they are still classified into the category of centralized Botnet.

2.1 Centralized Botnet

IRC-based Botnet: In the early days of the Internet, the earliest centralized Botnets were mainly IRC-based Botnets, which mainly used IRC services to communicate between C&C servers and Bots (**Figure 1(a)**). This type of Botnet has a simple structure and adopts the known plaintext protocol [11]. Through the monitoring of activity cycle of the Botnet (such as ports and messages), the characteristics can be clearly identified, and these data flow can be easily filtered out in the

Type	Protocol	Examples
Centralized	IRC-based Botnet	SdBot, AgoBot, GT-Bot, RBot
	HTTP-based Botnet	Rustock, ClickBot, Naz, Zeus, Conficker, Torpig
	Custom protocol Botnet	Mega D, Mariposa
Distributed	Structured P2P Botnet	PhatBot
	Unstructured P2P Botnet	Sinit, Nugache
	Hierarchical Botnet	Waledac, Storm

Table 1.
Classifications of some known Botnets.

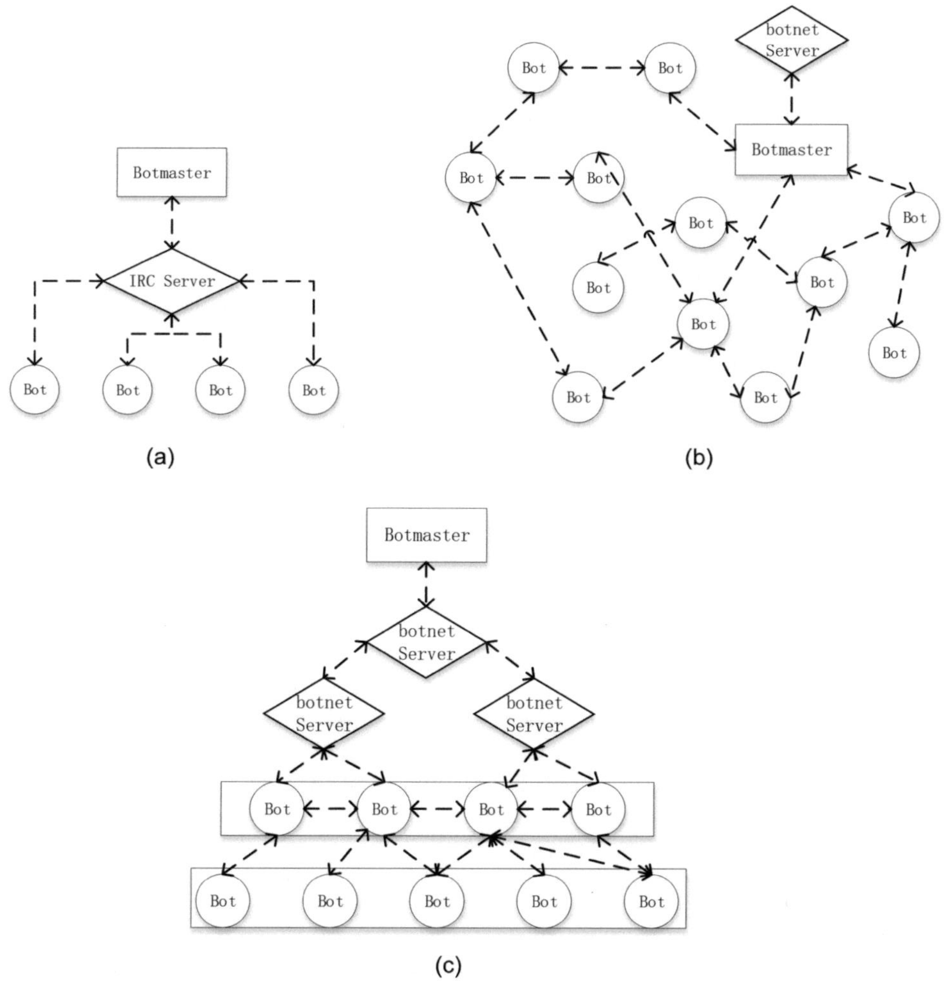

Figure 1.
Three types of Botnet structure.

network defense. This type of Botnet has a little impact because of its small scale. However, due to its simple operating mechanism and strong operability, it is deeply used by hackers. With the current development of Botnet, many hackers still use it.

HTTP-based Botnet: Due to the easy identification of messages of IRC-based Botnet, the HTTP-based Botnet arose. This type of Botnet could hide itself well by adopting HTTP protocol. Since the communication protocols between devices on the Internet are mainly HTTP protocol, HTTP messages in the information transmission of HTTP Botnet can be mixed with normal messages, making it difficult to filter directly through the router rules (ACL), which greatly improves the survival ability of Botnet and makes it more concealable. It is known that the HTTP-based Botnet is more complex and diverse than IRC-based Botnet. Rustock, Zeus, Torpig, etc. encrypt the content of the communication, and Conficker and Torpig also adopt a technique named "domain-flux" to increase the difficulty of blocking their control servers [12]. In addition, a small number of Botnets, such as Naz, also directly use popular social networking sites (such as Facebook, QQ space, etc.) as control servers, increasing the difficulty of detection and blocking [13]. Most Botnets currently use the HTTP protocol.

Custom protocol Botnet: Some Botnets use custom protocols for communication. The known Botnets of this type include Mega D, Mariposa, etc. Since Mega D uses a custom protocol, the first thing for researchers is to understand its operating mechanism through means of data mining and analysis or reverse capability. Compared with the IRC protocol and the HTTP protocol, Mariposa uses the UDP protocol for transmission, which does not require a three-way handshake. It is more difficult to be shielded by router rules (ACL), and its survivability is stronger.

2.2 Distributed Botnet

For the Botnets described above, the overall structure is a C&C server connected to multiple infected Bots. When the C&C server is broken by security experts, the Botnet is not available anymore. In order to enhance the survivability of Botnets, hackers increase the number of C&C servers and allow Bots to communicate with each other, so the distributed Botnet arise. This type of Botnet has a complicated structure, is difficult to construct, and requires a hacker with strong capabilities. At present, there are many distributed Botnets (such as Waledac and Storm), whose viability has been verified.

Structured P2P Botnet: The communication protocol between such Botnets is not unstructured (P2P protocol). A typical example of structured Botnet is PhatBot, which uses a fully connected Waste Protocol, which leads to a poor scalability of the PhatBot [14]. Early Storm adopted Overnet based on the Kademlia protocol [15] as a way of command and control. Since the information of other nodes can be obtained by the lookup operation in the Kademlia protocol, the researchers could make use of this feature to display the set points in all Overnet networks and then fill in many virtual set points (which we set), so that many messages and file transfers in Botnet will be introduced to the masquerading set points. In this way, the Bots are identified, and the judgment on the scale of Storm Botnet and the defense against it are finally achieved.

Unstructured P2P Botnet: The Bots under this model are connected irregularly, and they can communicate with each other. The communication method is also irregular, and they can send messages in a one-to-many way. There are many types of unstructured P2P Botnets, with two main ones (Nugache and Sinit). The operating mechanism of Sinit is random scanning, which adopts a scan code in the source code to filter some necessary IP segments, aimlessly identify other Bots. The message is sent through port 53, with a poor degree of concealment. The Nugache Botnet keeps a list internally. When the Botnet asks for a connection, it selects an uncertain record from the list of connection. If it is not successful, the random selection will continue; if it is successful, the connecting parties will refresh the list with each other [16]. Dittrich made an effort to keep sending message requests, refresh the list, and enumerate the whole Nugache network by recording and finally draws the structure diagram as shown in **Figure 1(b)**. It is found from the structure diagram that Nugache applies a range interval to the exit and entry massage of the Bot, giving birth to a P2P network with random connectivity. This decentralized topology, combined with the encryption of communications, allows Nugache to have very good concealment and keep a substantial number of active Bots unnoticed for a long time.

Hierarchical Botnet: This type of Botnet is referred to as hybrid P2P Botnet in some literature [17], and it is believed that the most prominent feature is the hierarchical structure. The structure is divided into at least three layers, the Bottom layer is the Bots, the middle layer consists of some Bots or C&C servers with better performance as the medium for information transmission, and the top layer is the core C&C server. This structure can prevent the top layer from being discovered by researchers and achieve more complex functions. Kanich et al.'s further research on Storm found that the Storm is a three-layer Botnet [18]. The Bots in the bottom layer could send HTTP messages, virus information, etc. The Bots can use the Internet to query other proxies infected host, and the most top hacker server (C&C server) is behind the proxy infected host, with a high degree of concealment.

Waledac is another large-scale hierarchical Botnet, which is also used to send large amounts of spam. Waledac has a similar hierarchical structure as shown in the above (**Figure 1(c)**). It has a structure of one more layer than the ordinary hierarchical structure, and relevant research shows that it is transformed on the basis of the previous hierarchical Botnet. Botmaster is mainly divided into four layers of institutions (from bottom to top for Spammer, Repeater, TSL, UTS). The lower two layers are computer devices with vulnerabilities. The upper two layers are the hierarchical C&C servers used by hackers. The communication method of this Botnet is a technology named fast-flux. The third layer (Repeater) serves as a bridge between the second layer and the fourth layer of Bots, that is, using Bot as a proxy. This is different from the Koobface [19] Botnet, which uses trusted social networking sites, game sites, and other large server devices as its own proxy layer. Waledac is more viable in this way. Nunnery et al.'s research found that Waledac is able to offer two different levels of spam business. Through experiments, the researchers found that due to the diversity of the Bots in the bottom layer of Waledac, it has the function of sending spam, but this ability is not strong, and it is easy to be directly intercepted by some large-scale defense servers; there is also a spam service that can be sent directly by the second layer (TSL) of the Waledac Botnet. This method of sending can dynamically modify the contents of the file to prevent it from being killed by the fixed antivirus software, with high availability. At the same time, in order to further improve the concealment of Botnet and prevent it from being detected by network supervisors, Waledac's internal message transmission mechanism is based on elliptic curve encryption to implement encryption technology. A two-in-one technique (timestamp + public key) is used on the communication between the second and third layers to prevent replay and forgery [20]. In order to prevent security personnel from tracking Botnet, Waledac adopts the detection method of domain name polling to prevent the population of fake nodes [21].

Koobface also adopts an intermediate node as a proxy to hide the control server. But Koobface is notable not for its complex structure but for its numerous functional modules and the way it uses social networking sites to spread its messages. Koobface steals the accounts of social networking sites on Bots, automatically logs in and sends malicious links to friends for transmission, which exploits the trust between social network users. Koobface has a range of modules targeted at almost all major social networks and can force infected users to recognize Captcha images, as well as DNS hijacking, search, hijacking, web server and information theft, etc.

3. Working mechanism of Botnet

As shown in **Figure 2**, the life cycle of a Botnet is divided into six phases: (1) There are many ways for a Botnet to propagate a Bot program, such as page virus, vulnerability attack, email phishing, etc.; (2) If the host is infected, then the Bot program will remain in the system; (3) hosts with vulnerabilities send domain name query to domain name server to obtain IP address of Botnet controller; (4) host with vulnerabilities will connect the Botnet controller and join Botnet; (5) communication connection between Bot and Botnet controller start, as well as the issuance and transmission of commands between attacker and Botnet controller; and (6) the Bot attacks the victim at the command given to the controller.

In phase 1, Botnet adopts email phishing or URL hidden connections to link to some web pages and runs malicious code on the page; this propagation mode is similar to worm propagation mode. Both of them are to attack vulnerable services

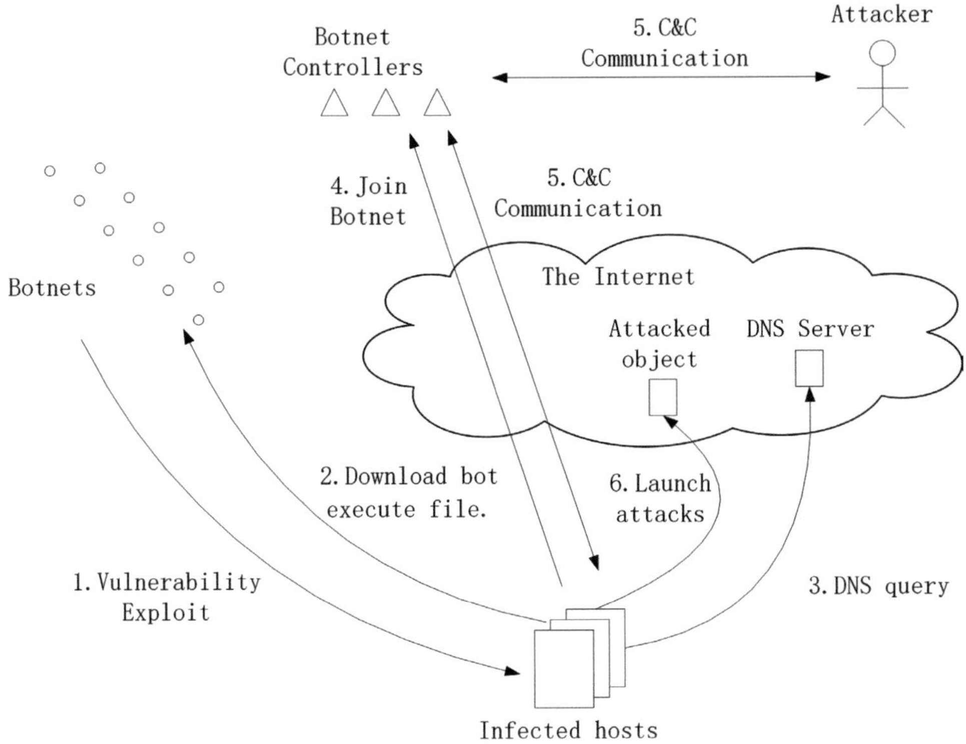

Figure 2.
Working process of Botnet.

by scanning specific ports with specific algorithms, which is very common. There are various algorithms to determine how and when to scan. The Bot does not implement any propagation at first until it receives the command from the attacker, which makes detection more difficult for Botnet [22].

In phase 2, when the computer with vulnerabilities is attacked, it will turn into Bot, and the C&C server will give a command of program installation (such as the echo command in Mirai). This process can be a one-step or multistep installation. For example, a control program is downloaded first, and then the entire Botnet program will be downloaded at a later stage. In addition, some Botnets that exist with chat software will also spread by Relay Node, which is not easy to be found, but also have problems such as delay.

In phase 3, the IP address of the early Botnet controller is directly written in the Bot program, which has the disadvantage of low concealment, so at this stage, the Bot program contacts the C2 controller through the DNS domain name.

In phase 4, because the victim host joined Botnet in different ways, in order to improve the security of the Botnet, it adopts a certain authentication mechanism. Only authenticated hosts can join the Botnet group and carry out communication and control interactions. In addition, the Botnet controller is also selected by the hacker in the Botnet group. In order to prevent these controllers from being shut down or offline, the attacker will generally adopt DNS technology to replace the domain name with a new IP address when the controller goes offline or it is captured. Furthermore, fast-flux technology is used to provide an IP list, and the IP address is periodically bound from the list to the domain name to improve reliability and detection difficulty. The Botnet also replaces the legitimate domain name server on the infected host with its own DNS name server, which has three benefits: (1) if the Bot program is cleared by the host user, some Bots will even reinfect the host through their own DNS name server; (2) make some antivirus programs unable to update itself; and (3) implement phishing attacks to enable users to access fake websites [23].

In phase 5, the main activity is C&C communication, receiving information sent by the hacker. Botnet maintains communication with the Bot and at the same time protects itself from being captured by the security system. Bot will accept or actively acquire commands, infect more machines, or download updates to the Botnet code. At this stage, due to the original fixed IP, fixed domain name, dynamic update, etc. are less concealed, and Botnet will often adopt domain-flux or fast-flux technology to improve its survivability.

Domain-flux technology is created to solve the problem of central point failure. The attacker uses the domain-flux protocol to prevent itself from shutting down by the defense personnel. The C&C domain name accessed by the Bot is no longer statically hard coded but can be dynamically generated, which allows the C&C server to communicate securely with the Bot [24]. The principle of the domain name algorithm is DGA algorithm, which puts a comprehensive factor such as a dictionary, a random number, a date, and a hot topic into a generation algorithm, generates a string of special character prefixes, and adds a TLD to obtain a final domain name resource. Because of its fast generation speed and high frequency, even with the use of blocking, shielding, and other measures, it cannot protect against invasion. Torpig and Conficker, which appear on the web in general, adopt this technical feature. At the beginning of the twenty-first century, the foundation of fast-flux appeared and gradually attracted more and more attention. Fast-flux is created to address the problem of security personnel locating C&C server domains and IP (both bound to each other) through reverse technology. In general, when a domain name server is used to query the IP of a certain domain name, the result of the query will return the same IP in a short period of time because of the DNS cache. However, fast-flux technology can constantly change the correspondence between IP addresses and domain names, and it makes a large number of queries in a short period of time to return to different results. The fast-flux is divided into two categories (single-flux and double-flux) according to the different number of mapping layers. Single-flux is the fast-flux that has only one mapping layer, a domain name that has one and only one continuously changing IP address. Double-flux represents the fast-flux with two mapping layers. In the actual Internet environment, hackers deploy multiple domain name servers. By modifying the domain name of the top-level server, the correspondence between the IP address of the lower-layer DNS server and the domain name is constantly changing. A Botnet employs fast-flux technology, which would have a large number of C&C servers, and most of the servers are not controlled by the hackers themselves but by Bots. During the check, the security personnel will find that there is no control command from a hacker on the "C&C servers"; these controllers are only responsible for the command forwarding and springboard function, which virtually improves the concealment of the Botnet. Fast-flux technology can also be used to break the domain names of certain phishing websites and malicious websites. Storm adopts this technology to analyze the domain name that sends the message. Phish rock criminal organization adopts it to resolve the domain name of phishing website [25]. Waledac also adopts fast-flux technology to conceal its control server.

In phase 6, the Botnet receives the command sent by the hacker and launches the attack. The attack modes (as shown in **Table 2**) are different [26]; the number of Bots participating in the attack, attack target, and the attack means can also be completely controlled by the hacker. Botnet initially launches a single- or multi-machine distributed denial-of-service attack. Gradually, Botnet turns into profitable attacks, such as stealing users' privacy information on victim machines. For many years, Symantec's global annual cybersecurity report stated that the vast majority of spam is sent by Botnet. Spam sent by Botnet is more harmful than

Attack mode	Difficulty for detection	Complexity	Damage
Small-scale DDoS attack	High	Low	Low
Large-scale DDoS attack	Medium	Medium	High
Stealing information	Low	High	Medium
Sending spam	Medium	Medium	High
Phishing	Medium	High	Medium

Table 2.
Common modes and characteristics of attack initiated by Botnet.

regular spam, making detection more difficult. The process of phishing attack is initiated by Botnet: the Bot erases and replaces the addresses of legitimate DNS on the machine. When the user accesses the confidential page, the replaced domain name server sends the phishing website page to the user [27].

4. Botnet threats and assessment

The threat assessment of traditional Botnet mainly starts from its several key performances; the stronger the key performance of Botnet, the stronger the threat. The key performance indicators of traditional Botnet mainly include four points: transparency, concealment, destruction resistance, and attack capacity.

The transparency of Botnet is mainly reflected in that when an attacker maintains a Botnet or orders a Botnet to attack a certain site, the Botnet can be operated as a whole and there is no need to pay any attention on the internal details of the Botnet. This transparency is mainly realized through the control structure. Attackers input operation commands and control information into the control structure, and the control structure continuously transmits relevant contents to various nodes, so as to control the Botnet as a whole.

The concealment of Botnet means that the activities in the main stages of the life cycle of traditional Botnet need to be carried out covertly, to effectively reduce the possibility of detection of the nodes, operating facilities and overall data flow of Botnet, etc. The concealment of Botnet requires that network nodes should not occupy memory and broadband resources too significantly and the damage to the availability of controlled hosts should be relatively small. The most important thing is to prevent itself from checking by the end user to avoid being discovered by the network security supervision system.

The destruction resistance of Botnet mainly refers to the key characteristic that Botnet is able to maintain its attack ability when some nodes are cleared or destroyed, which is also called tenacity. The great performance of destruction resistance makes the Botnet have strong survivability and can create more superior conditions for the attacker to adjust the behavior characteristics of the Botnet node, thereby effectively avoiding the occurrence of the entire Botnet failure. The main way is to build a more robust structure of Botnet to improve its destruction resistance.

The attack capacity of Botnet mainly refers to the sum of all controllable resources that can be controlled by an attacker. The attack capacity determines the maximum attack strength that an attacker can initiate, and the attack capacity mainly depends on broadband resources and network size. The attack flow that an attacker can initiate increases with the increase of broadband resources. The larger the network size, the more URLs can be exploited by an attacker, and the more dispersed attack source, the fewer constraints in the attack process.

These key performance indicators can be roughly divided into three categories: transparency and concealment belong to the Botnet's defense capability, destruction resistance belongs to the Botnet's survivability, and the attack capacity belongs to the Botnet's attack capability. In addition to the above key performance indicators, there are some more detailed indicators, but they fall within these three capabilities, such as command accessibility of Botnet, node averaging of Botnet, Botnet resilience, etc.

With the rise of the Internet of Things, the rapid development of smart terminals, and the continuous improvement of mobile network technologies, in addition to traditional Botnet, mobile Botnet has become one of the main platforms threatening mobile network security. After the mobile Botnet invades the intelligent terminals in the mobile Internet, these smart terminals are controlled in a one-to-many way through controlling and command channels. It can be seen that mobile Botnet is a subset of traditional Botnet, but it is far more harmful to users than traditional Botnet. Due to the particularity of the mobile network, its threat assessment has its own unique indicators in addition to the key performance indicators of traditional Botnet. The threat assessment for mobile Botnet can be started with the following performance indicators: attack performance, defensive performance, survivability, auxiliary performance, and environmental performance. There are more specific indicators in each performance indicator, such as confidentiality and node control efficiency in attack performance, stability and anti-detection capability in defense capability, network averaging and network connectivity in survivability, propagation capabilities and command mechanism performance in auxiliary performance, scalability and loan consumption in environmental performance, and more.

5. Conclusions

At present, various cyberattacks based on Botnet are the most serious security threats to the Internet. As Botnet continue to evolve and behavioral research on Botnet is inadequate, the question of how to apply some behavioral problems to Botnet research and combine the psychology of the operator to analyze the future trend of Botnet is still a continuous and challenging issue.

Botnet is a common computing platform which can be controlled remotely by attackers by invading several noncooperative user terminals in the network space. It is an attacking platform consisting of multiple Bots controlled by a hacker. The behavior of Botnet is also controlled by the hacker, rather than being controlled by certain code logic, which also makes it difficult to locate and predict the Botnet attack. The Botnet is developed in two phases: it was the primary virus and worm in the first phase, and it transformed into Botnet platform in the second phase. The virus attack has the characteristic of integration. Botnet is different, the control command of Botnet is issued by separate C&C server, and the attack and invasion are completed by the controlled hosts.

Botnet has many types of classification, and it can be divided into centralized Botnet and distributed Botnet according to different operating principles. The difference is that there is only one C&C server in the entire network platform for the centralized Botnet, and the infected nodes also communicate with each other in the distributed Botnet.

The attack process of the Botnet is mainly divided into six phases: in the first phase, Botnet will spread through various traditional viruses or worms; in the second phase, the Bot begins to download the entire Botnet program; in the third phase, the Bot contacts Botnet controller; in the fourth phase, the Bot is authenticated, and the authenticated Bot can join the Botnet group; in the fifth phase, C&C communication between Botnet and Bot will start to receive information sent by the hacker; and in the sixth phase, the Botnet launches an attack based on commands sent by the hacker.

The Botnet is popular all over the world, which poses a huge threat to the global Internet and the Internet of Things. DDoS attack is still one of the largest Internet security threats in the world, and the DDoS attacks are mainly launched by Botnet.

Acknowledgements

The authors thank Lu Yao, Qing Ye, Na Wang, Yicheng Lu, Haichang Yao, Kui Li, and Ruchuan Wang for their contributions. This work is supported by the National Key Research and Development Program of China (2017YFB1401301, 2017YFB1401302, 2017YFB0202204), the National Natural Science Foundation Program of China (61373017), the Key Research and Development Program of Jiangsu Province (BE2017166), the Natural Science Foundation Outstanding Youth Fund of Jiangsu Province (BK20170100), the Open Fund of Jiangsu High Technology Research Key Laboratory for Wireless Sensor Networks (WSNLBZY201514), and the 1311 Project of Nanjing University of Posts and Telecommunications.

Conflict of interest

The authors declare no conflict of interest.

Author details

Ji Yimu[1,2,3,4,5] and Liu Shangdong[1,3,4*]

1 School of Computer Science, Nanjing University of Posts and Telecommunications, Nanjing, China

2 Jiangsu High Technology Research Key Laboratory for Wireless Sensor Networks, Nanjing, Jiangsu, China

3 Institute of High-Performance Computing and Bigdata, Nanjing University of Posts and Telecommunications, Nanjing, Jiangsu, China

4 Nanjing Center of HPC China, Nanjing, Jiangsu, China

5 Jiangsu HPC and Intelligent Processing Engineer Research Center, Nanjing, Jiangsu, China

*Address all correspondence to: lsd@njupt.edu.cn

References

[1] China Internet Network Information Center. The 43rd Statistical Report on Internet Development in China. 2019;**2**:19-20. DOI: 10.13666/j.cnki.jnlc.2019.02.001

[2] Kolias C, Kambourakis G, Stavrou A, et al. DDoS in the IoT: Mirai and other Botnets. Computer. 2017;**50**(7):80-84. DOI: 10.1109/MC.2017.201

[3] Cai T, Zou F. Detecting HTTP Botnet with clustering network traffic. In: International Conference on Wireless Communications, Networking and Mobile Computing. Shanghai. IEEE. 2012. pp. 1-7. DOI: 10.1109/WiCOM.2012.6478491

[4] Song YZ. P2P Botnet detection based on permutation entropy and multi-sensor data fusion on decision level. Computer Science. 2016;**43**: 141-146. DOI: 10.1145/2379616.2379622

[5] Jianen Y, Zhaoxin Z, Haiyan XU, et al. Detection of IRC Botnet C&C channels using the instruction syntax. Journal of Tsinghua University. 2017;**57**(9):914-920. DOI: 10.16511/j. cnki. cnki.qhdxxb.2017.26.040

[6] Jang DI, Kim M, Jung HC, et al. Analysis of HTTP2P Botnet: Case study waledac. In: IEEE Malaysia International Conference on Communications. 2010. DOI: 10.1109/MICC.2009.5431541

[7] Dibenedetto S, Gadkari K, Diel N, et al. Fingerprinting custom Botnet protocol stacks. In: IEEE Secure Network Protocols. 2010. DOI: 10.1109/NPSEC.2010.5634448

[8] Yu-Peng T, Zhang YZ, Yin T. Modeling and evaluating a cross-realm architecture for P2P Botnet. Acta Electronica Sinica. 2018;**46**(4):791-796. DOI: 10.3969/j.issn.0372-2112.2018.04.004

[9] Wu Z, Zhou H, Yu Z. A novel hierarchical Botnet model. In: IEEE Conference Anthology. China. 2014. DOI: 10.1109/ANTHOLOGY.2013.6784723

[10] Sinha P, Boukhtouta A, Belarde VH, et al. Insights from the analysis of the mariposa Botnet. In: International Conference on Risks & Security of Internet & Systems. Montreal, QC, Canada. 2010. DOI: 10.1109/CRISIS.2010.5764915

[11] Wang T, Yu SZ. Centralized Botnet detection by traffic aggregation. In: IEEE International Symposium on Parallel & Distributed Processing with Applications. Chengdu, Sichuan, China. IEEE; 2009. DOI: 10.1109/ISPA.2009.74

[12] Bonneau J. The Science of Guessing: Analyzing an Anonymized Corpus of 70 Million Passwords. Security & Privacy. San Francisco, CA, USA. IEEE; 2012. DOI: 10.1109/SP.2012.49

[13] Mazurczyk W, Caviglione L. Steganography in modern smartphones and mitigation techniques. IEEE Communication Surveys and Tutorials. 2015;**17**(1):334-357. DOI: 10.1109/COMST.2014.2350994

[14] Barford P, Yegneswaran V. An Inside Look at Botnets. Advances in Information Security. 2007;**33**(4): 171-191. DOI: 10.1007/978-0-387-44599-1_8

[15] Li Z, Chen X. Misusing Kademlia protocol to perform DDoS attacks. In: IEEE International Symposium on Parallel & Distributed Processing with Applications. Sydney, NSW, Australia. 2008. DOI: 10.1109/ISPA.2008.15

[16] Singh K, Guntuku SC, Thakur A, et al. Big data analytics framework for peer-to-peer Botnet detection using random forests. Information Sciences.

2014;**278**:488-497. DOI: 10.1016/j.ins.2014.03.066

[17] Abaid Z, Kaafar MA, Jha S. Early detection of in-the-wild Botnet attacks by exploiting network communication uniformity: An empirical study. In: Ifip Networking Conference. Stockholm, Sweden. 2018. DOI: 10.23919/IFIPNetworking.2017.8264866

[18] Kanich C, Kreibich C, Levchenko K, et al. Spamalytics: An empirical analysis of spam marketing conversion. Communications of the ACM. 2009;**52**(9):99-107. DOI: 10.1145/1562164.1562190

[19] Thomas K, Nicol DM. The Koobface Botnet and the rise of social malware. In: International Conference on Malicious & Unwanted Software. Nancy, Lorraine, France. 2011. DOI: 10.1109/MALWARE.2010.5665793

[20] Rossow C, Andriesse D, Werner T, et al. SoK: P2PWNED - Modeling and evaluating the resilience of peer-to-peer Botnets. In: IEEE Symposium on Security & Privacy. Berkeley, CA, USA. 2013. pp. 97-111. DOI: 10.1109/sp.2013.17

[21] Stringhini G, Hohlfeld O, Kruegel C, et al. The harvester, the botmaster, and the spammer: On the relations between the different actors in the spam landscape. In: Acm Symposium on Information. ACM; 2014. DOI: 10.1145/2590296.2590302

[22] Li Z, Goyal A, Yan C. Honeynet-based Botnet scan traffic analysis. Botnet Detection. 2008. DOI: 10.1007/978-0-387-68768-1_2

[23] Peng T, Harris I, Sawa Y. Detecting phishing attacks using natural language processing and machine learning. In: 2018 IEEE 12th International Conference on Semantic Computing (ICSC). IEEE; Laguna Hills, CA, USA. 2018. DOI: 10.1109/ICSC.2018.00056

[24] Yadav S, Reddy AKK, Reddy ALN, et al. Detecting algorithmically generated domain-flux attacks with DNS traffic analysis. IEEE/ACM Transactions on Networking. 2012;**20**(5):1663-1677. DOI: 10.1109/TNET.2012.2184552

[25] Cook DL, Gurbani VK, Daniluk M. Phishwish: A simple and stateless phishing filter. Security & Communication Networks. 2010;**2**(1):29-43. DOI: 10.1002/sec.45

[26] Khattak S, Ramay NR, Khan KR, et al. A taxonomy of Botnet behavior, detection, and Defense. IEEE Communication Surveys and Tutorials. 2014;**16**(2):898-924. DOI: 10.1109/SURV.2013.091213.00134

[27] Tanner BK, Warner G, Stern H. Koobface: The Evolution of the Social Botnet. Ecrime Researchers Summit. 2010. pp. 1-10. DOI: 10.1109/ecrime.2010.5706694

Chapter 4

Evaluation of Botnet Threats Based on Evidence Chain

Liu Shangdong and Ji Yimu

Abstract

The current network security faces a serious threat, which has been brought about by the large-scale proliferation of botnet, and its detection has become one of the important tasks of the existing cyberspace security. At present, although network administrators have firewalls, intrusion detection systems, intrusion prevention systems, and other technical means to achieve partial network protection, they are still confronted with severe challenges in the detection and prevention of a botnet known as a threatening attack platform. The new botnet is characterized by its large scale and multifunction. Further, it is hard to detect, and it may cause a sharp decline in the normal defense level of the protected object in a short period of time. In this chapter, we propose a method of botnet threat assessment based on evidence chain. The DS evidence theory is used for network security situational awareness. On the basis of determining the recognition framework, all possible results are considered, and each evidence is assigned a basic credibility, and the final credibility of the target is fused by using the composition rule. The experiments show that this method can work efficiently and detect the major threats in the protected network in time.

Keywords: botnet, intrusion detection, situational awareness, evidence chain, threat evaluation

1. Introduction

In recent years, with the rapid development of Internet of Things (IOT) technology, more and more devices are exposed to the Internet. These devices are complex in variety and explosive in number. This kind of interconnected environment will make the security risk increase and spread rapidly, and bring severe security problems. Among all kinds of security problems, botnet in particular brings serious harm. Botnets are made up of "zombie hosts" infected with a malicious code that infect normal devices, forming a large-scale "botnet" of IOT, once the "botnet" launches a distributed denial of service attack. This will wreak havoc on the Internet infrastructure [1].

In view of the large scale of botnet, the variety and number of botnet hosts, and the unpredictable vulnerability types, the network security protection should be considered from the overall situation. Therefore, it is very important to grasp the information of the network and to perceive the status and development trend of the network security. Network situational awareness can capture the security elements that cause the change of network situation in a large-scale network environment, and make decisions and actions by acquiring, understanding, predicting, and making decisions [1]. The concept of Situational Awareness (SA) originates from the military demand in the 1980s, and with the rise of network, it was introduced by Tim Bass into the field of network security.

SA should go through several steps, such as situation acquisition, situation understanding, situation prediction, situation visualization and so on [2, 3]. In the situation acquisition stage, there may be a lot of complex, repetitive, or even false alarm information. In addition, the existing SA methods use IDS, firewalls, virus detection and other tools data, based on time series, graph theory, Bayes, game theory and other methods, according to the network environment, the history of the attacker and the network ontology vulnerability; these are used to evaluate and predict the network security situation, without considering the emerging vulnerabilities and their SA.

To solve the above problems, this chapter proposes a botnet SA method based on DS evidence theory. Compared with other SA methods, DS evidence theory not only can solve uncertainty problems, but it also does not need prior probability and conditional probability density. Therefore, we can manually assign it initial trust based on our expertise and individual knowledge.

Botnet SA integrates all kinds of botnet security elements to evaluate the security situation of the network in real time, which provides the basis for the network security analysis, and evaluates the network security more accurately, thus minimizing risks and losses from botnet threats. Botnet security SA plays an important role in improving the ability of network monitoring, emergency response and predicting the development trend of network security.

The main contributions of this chapter are as follows:

1. we propose a method of botnet threat assessment based on evidence chain, which computes the target credibility to determine whether there is a threat in the network;

2. the evidence chain method is applied to botnet to realize the situation of network security. DS evidence theory solves the uncertainty problem of network threat.

3. the experiment is carried out using the public data set of Nanjing University of Posts and Telecommunications (NJUPT). The results show that the network security situation assessment method proposed in this chapter is reasonable and effective, and can improve the accuracy of security situation prediction.

2. Related work

There are already some approaches to network security SA: In the research of network security SA architecture, Kokkonen proposed in 2016 a network security SA architecture, which mainly includes information exchange module and emphasizes standardized information format [4]. In 2017, Eiseler proposed a network security SA architecture from the perspective of IT complexity [5]. The main idea is to abstract a layer of operation (decision) and the result of decision for decision makers from non-technical background. In the research of network security SA, in 2016, Yang et al. used SVM machine learning method for SA [6]. After being trained by classifier, the data can be used to predict the situation value. But the method has the defect that the situation is normalized and the information is not abundant enough. In 2016, KHALID et al., targeting data injection attacks, could lead to unreliability and insecurity of network physical infrastructure such as (WAMS), a wide-area monitoring system. In this chapter, a Bayesian based approximate filter (BAF) method [7] is proposed to minimize the impact of injection attack on oscillatory parameters, so as to improve the resistance of monitoring applications to data injection attacks. In 2016, in the HMM-based network security situation assessment method, Li et al. used to extract the observation values and model parameters by establishing the time period, which is an important factor affecting the real-time and accuracy of the evaluation. However, there are two problems: The results are as follows: (1) the size of the time period is given randomly by people, which cannot represent the security and real-time performance of the current network; (2) the state transition matrix and the observation symbol matrix are usually determined by experience and have strong abstractness. To solve this problem, Li et al. later trained the parameters of the HMM model by mixed multi-population genetic algorithm (MPGA) [8] to improve the reliability of the parameters and to solve the problem that the emergency situation could not be highlighted in a certain period of time. Experiments show that this method can reflect the current network security situation effectively and accurately. [9, 10] put forward the overall goal of network security SA, which is determined by scope, level, requirement and decision. The method of SA is classified from four aspects: data collection, decision making, analysis and visualization.

Through the research of network security SA, to a certain extent, the researchers give other researchers some practical methods, but these methods also have a limited scope of application. Most of the SA methods only consider the calculation of the threat situation caused by an external attack and ignore the problem of the security situation change caused by the insecurity of the system and the equipment itself. This chapter presents a method of network security SA based on evidence chain theory. DS evidence chain theory has many advantages in SA. Firstly, it does not require prior probability and conditional probability density. Secondly, sometimes the information provided by the sensor is not necessarily very accurate, and there may be a certain degree of fuzziness, and the DS evidence method can solve the uncertainty calculation problem. Finally, DS evidence theory can continuously narrow the scope of the hypothesis set by merging evidence. Its basic idea is to fuse several sub-evidences according to the Dempster formula, so as to further determine the possibility of the occurrence of certain propositions.

3. A method of network SA awareness based on evidence chain

The chain of evidence is a collection of evidence formed by two or more evidence links connected by the chain heads for a certain object of proof. Due to the complexity of the current network environment and the emergence of various network attack methods, the management requirements and the means of recording technology are different. The vulnerabilities of most network and system are scattered and independent, and the performance cannot fully reflect the real situation of the network status. It needs to combine the vulnerabilities and network status transformation together through the relevance of vulnerabilities, and to connect them according to the inherent meanings and logical relationships to form a chain structure that is mutually connected and mutually validated, which involves the chain of evidence for network situation awareness.

3.1 The components of the chain of evidence

The components of chain of evidence for audit include chain link, chain connection and chain domain. Among them, chain link refers to the single evidence that constitutes the chain of evidence for situation awareness, also known as the node evidence, which is expressed as a single physical object; chain connection is an overlapping or embedding relationship or logical reasoning relationship between the single evidences; chain domain refers to the entire information set (all evidences) that the auditing entity can understand or know when verifying a certain network activity under the existing cognitive ability and technical conditions. The scope of chain domain is determined by the network activity and the cognitive ability of the network entity, and the maximum value is all the facts required for situation awareness, and the minimum is the main facts of situation awareness.

3.2 The essence and attributes of the chain of evidence

The evidence for situation awareness is essentially the retention of information about the past network activity of the object, and the retention of such information is the record and reflection of the network activity which objectively exists. When these records and reflections do not fully capture the main facts of a network activity, it needs to be achieved by constructing a chain of evidence. Therefore, the essence of the chain of evidence for situation awareness is that different evidences of different segments or conditions of the same network activity, through the multi-component chain-dependent relationship in terms of meaning and logic, mutually confirm each other and connect with each other to jointly reveal the truth of the same economic activity. The chain of evidence for situation awareness not only has the characteristics of the adequacy and appropriateness of general audit evidence, but also has the characteristics of relevance, integrity and complexity, etc. of unique or different meanings. Among them, relevance refers to the objective connection of causal relationship, conditional relationship and space–time relationship between the evidences of each link constituting the chain of evidence. Integrity means that the evidences of each link constituting the chain of evidence have a consistent proof effect and proof direction, and together constitute a complete proof system. Complexity refers to the complex source of evidence of each link that constitutes the chain of evidence. There are some evidences from the same source, that is to say they come from the same network activity; and there are some evidences from different sources, but the contents of them are involved each other. The evidences of each link sometimes have different forms, and the evidence of entity coexists with the evidence of person. The contents of the evidences are coherent and overlapping, and there is other information unrelated to the audit findings.

3.3 Connection mode of chain of evidence

The chain of evidence can be divided into two kinds of connections, explicit and implicit, according to whether there are semantic intersections and overlapping relationships between links, such as explicit texts. Among them, explicit connection refers to the overlapping and embedding of evidences contents between adjacent business processes in the chain of evidence. Implicit connection refers to the connection relationship between evidences formed by logical reasoning. Node evidence in the chain of evidence can be divided into core evidence and auxiliary evidence according to their different proof functions in network activities. Among them, core evidence, also called direct evidence, refers to the evidence that plays a major role in proving the emergence and existence of witnessed network activities. Auxiliary evidence is the evidence supporting core evidence, including making up the quality defects of core evidence and enhancing the persuasiveness of core evidence. The composition of the chain of evidence is shown in **Figure 1**.

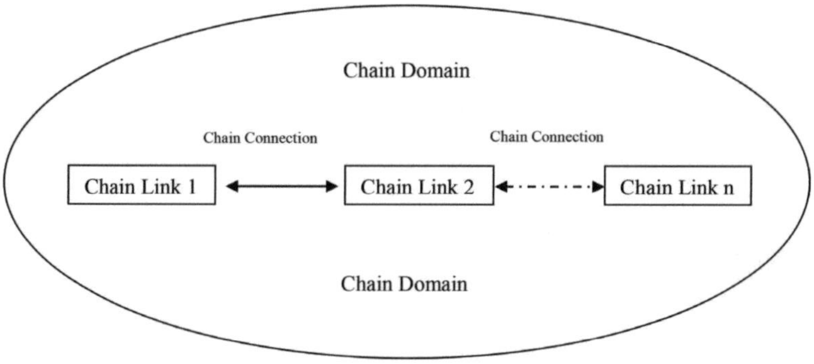

Figure 1.
Composition of chain of evidence.

3.4 Basic concepts of evidence theory

D-S evidence theory [11, 12] adopts mathematical reasoning to perform fusion calculations of inexact and incomplete information. In the D-S evidence theory fusion algorithm, the recognition framework is the framework of the whole judgment; the Basic Probability Allocation is the basis of fusion; the combinational rule is the fusion process, and the trust function and likelihood function are used to express the upper and lower limits of support strength interval of fusion conclusion to a hypothesis.

1. Recognition framework

Θ is a mutually exclusive non-empty finite set, which is known as recognition framework. It consists of N nonintersecting sets of $w_1, w_2, w_3. \dots w_N$, and there are N possible hypotheses in this recognition framework. The task of the evidence theory fusion algorithm is to estimate the trust level to each possible hypothesis.

2. Basic probability allocation

Basic Probability Allocation (BPA) is a function known as E.g. (1)m function. $m : 2^{\Theta} \rightarrow [0, 1]$,

And it satisfied:

$$m(\Phi) = 0; \sum_{A \subseteq \Phi} m(A) = 1 \tag{1}$$

When an evidence is constructed, each possible hypothesis or hypothesis combination within the recognition framework should be assigned with a trust level between [0, 1], and the sum of the trust levels of all hypotheses or hypothetical combinations should equal 1.

3. Trust function

The fusion conclusion of D-S evidence theory expresses the support strength for any hypothesis through an interval, and the lower limit of this interval is called the trust function, and the trust function is also called the Belief Function (bel). The trust function is defined in the recognition framework Θ as is Eq. 3:

$$bel(A) = \sum_{B \subseteq A} m(B)(\forall\, A \subseteq \Theta) \tag{2}$$

The trust function of a hypothesis in the fusion conclusion only calculates the support strength for the hypothesis directly during the fusion calculation, and does not calculate the support strength for the combination containing the hypothesis. If a part of the support strength in the Basic Probability Allocation is assigned to an unknown domain, then the support strength of this part cannot be calculated in the trust function.

4. Likelihood function

The upper limit of the fusion conclusion interval of D-S evidence theory is called the likelihood function, and the likelihood function is also called the Plausibility Function (pl). The likelihood function is defined in the recognition framework Θ as is Eq. 3:

$$pl(A) = \sum_{B \cap A \neq \Phi} m(B) = 1 - bel(\overline{A}) \tag{3}$$

The likelihood function of a hypothesis in the fusion conclusion not only calculates the support strength for the hypothesis directly during the fusion calculation, but also calculates the support strength for the combination containing the hypothesis and the support strength allocated to an unknown domain. The fusion conclusion could directly adopt trust function, likelihood function, even the interval formed by the trust function and likelihood function to express the support strength for each possible hypothesis.

5. Dempster's combinational rule

The Dempster's combinational rule, also known as the evidence combination formula, can be expressed as Eq. 4:

$$\begin{aligned} \mathrm{m}(A) &= m_1(A_1) \oplus \mathrm{m}_2(A_2) \oplus \mathrm{m}_3(A_3) \oplus \cdots \oplus \mathrm{m}_n(A_n) \\ &= \frac{1}{1-k} \sum_{A_1 \cap A_2 2 \cap \cdots \cap A_n = A} \prod_{i=1}^{n} \mathrm{m_i A_i} \qquad (\forall A \subseteq \Theta) \end{aligned} \tag{4}$$

Where k is the degree of conflict of evidence, $\frac{1}{1-k}$, $k = \sum_{A_1 \cap A_2 \cap A_3 \cap \cdots A_n \neq \Phi} m_1(A_1) \bullet m_2(A_2) \bullet \cdots \bullet m_n(A_n)$. k = 1, the conflict between the evidences is so great that the evidence cannot be fused using the Dempster formula. When some These, two characteristics of the D-S evidence theory combination rule facilitate us in the combination of evidence. When combining multiple evidences, it does not need to consider combination orders. At the meanwhile, when there are consistency and contradiction between the evidences, group similar evidence into groups and then carry out the combination of grouped combination conclusions.

3.5 Research on application of evidence theory

Evidence theory has been widely used in the fields of expert system, information fusion, intelligence analysis, target judgment, legal case analysis, multi-attribute decision analysis, etc. due to its extensive advantages in algorithm and application level. Many researchers have also carried out corresponding improvement research on the problems in the application. As far as the algorithm itself is concerned, there are three main aspects from the terms of application:

1. Construct a corresponding fast algorithm for a specific evidence organization structure

In different application fields, the organization structure and expression form of evidence are different. Starting from the evidence itself, it is an important point in the application field to study the algorithm that can quickly obtain the fusion conclusion in the application.

2. Approximate calculation

Aiming at the problem that the computation amount will increase rapidly when the dimension of evidence theory fusion algorithm and the quantity of evidence increase, the approximate algorithm is constructed starting from the practical application. The method of approximate calculation can simplify the calculation process under the condition of ensuring the calculation conclusion of uncertain reasoning.

The basic idea of approximate calculation is to reduce the number of focal elements to achieve the purpose of reducing the amount of calculation.

Voorbraak found that if the combination of m functions will produce a Bayes trust function (i.e. a probability measure on a recognition framework), and then the substitution of m function with their Bayes approximation will not affect the result of Dempster's combinational rule, which is called the "Bayes" approximation method.

The meaning of the "Bayes approximation" is that it is very useful and computationally efficient for those cases where the final conclusion is concerned only with identifying the "elements" of the framework (i.e., a single hypothesis) rather than its "subset" (i.e., a subset of multiple hypotheses). Dubois and Prade proposed a "Consonant approximation" which is characterized by that the focal elements are nested after approximate calculation, and the number of focal elements does not exceed the number of hypotheses in the identification framework. The disadvantage is that this method is not suitable for calculation by Dempster's combinational rule, which may produce a large error. The "Consonant approximation" method applies to the expression of evidence.

Tessem proposed "(k, l, x) approximate algorithm ", k represents the minimum number of retained focal elements; l represents the maximum number of retained focal elements; x represents the maximum m value that is allowed to be deleted, and x usually takes a value on [0, 0.1].

First, sort the m value from big to small, and then loop the sum of m function values successively. If the number of retained focal elements is equal to 1, or the sum of the calculated m functions is greater than or equal to 1-x, the loop ends; otherwise, continue the loop, and finally normalize the m function values corresponding to the retained focal elements. The (k, l, x) method gives neither Bayes m function nor a consonant m function, but it does reduce the focal element.

3. Modification of D-S Method

In view of the problems existing in the practical application of D-S evidence theory fusion algorithm, corresponding modifications are made on the basis of traditional combination rules to avoid the irrationality of fusion conclusion under special circumstances.

4. Network security SA approach based on evidence chain

This section briefly introduces the flow of network SA [13] based on DS evidence theory: First, the identification framework should be determined, and all possible results should be considered, and each evidence should be assigned a basic credibility, and then the final credibility value of the target should be fused by using the composition rule. In this section, a method of SA based on DS evidence theory is proposed.

The network security SA based on DS evidence chain collects the protected network information through active and passive network sensors and takes the information as the fusion data of DS evidence theory after processing. Each piece of data collected by the sensor can be corresponding to one evidence, and then the corresponding initial credibility can be given to the evidence. Finally, the composite formula is used to fuse these evidences to obtain the credibility of the protected network threat proposition. This value reflects the degree of trustworthiness of the protected network under the threat of the evidence, and sets the confidence threshold. If the credibility exceeds the threshold, it indicates that the network component has a security threat and is vulnerable to attack, otherwise, the network component is secure.

In this chapter, the identification framework is $\Theta = \{T, F\}$ in which T indicates the camera was dangerous and vulnerable to attack while F indicates that the camera is secure and is not vulnerable to attack. Then the power set is $2^{\Theta} = \{\Phi, T, F, H\}$ in which Φ indicates the camera is both dangerous and safe while Himplies the camera may or may not be safe. The trust function satisfies $m(\Phi) + m(T) + m(F) + m(H) = 1$ in which $m(\Phi) = 0$ and $m(H) = 0$.

Second, every piece of data that is scanned from a camera device is used as a piece of evidence, and there are three types of evidence. The first is to scan the IOT devices opened on the port 23 all over the school, in which the camera device is the object of our SA so it could be attacked. An initial trust value is assigned to this evidence, that is, the ratio of camera devices to the number of devices opened on port 23 is used as the initial trust probability function of the evidence; the second type of evidence scans camera devices, in which cameras with weak password vulnerabilities are vulnerable to attack. Here we take the ratio of camera equipment with weak password vulnerability to the total number of cameras in NJUPT as the initial confidence probability function of the evidence; the third kind of evidence is to upload the virus to the camera device with weak password vulnerability. The successful uploading of the virus is highly dangerous and vulnerable to attack. We use the ratio of a successful webcam uploaded by a virus to a camera with a weak password vulnerability as the initial trust probability function. Through the above methods, we adopt three different types of evidence, further improve the credibility of evidence fusion, at the same time, we also compress a large number of evidence data into three pieces of evidence, improve the efficiency and time of synthesis. After that, we can use the improved composite formula to fuse the three evidences against the camera, and obtain the ultimate credibility of the dangerous situation of the camera in NJUPT.

Finally, the credibility $m(T)$ after fusion will be compared with a given threshold. If the reliability is greater than the threshold, it shows that the whole situation of the camera in NJUPT is dangerous and vulnerable to attack, otherwise, the overall situation of the camera of NJUPT is safe.

5. Experiment

In order to verify the feasibility and effectiveness of this method, the Telnet port scanning record of the network equipment in the campus network of NJUPT was used as the data source. The data was collected from the outbreak of a large-scale Mirai botnet attack on the East Coast of the United States at the end of 2016. The scope of collection is limited to the campus network of NJUPT. The study found that a large number of cameras in the campus network have weak password vulnerabilities. As shown in **Figure 2**, this vulnerability allows for intrusion into the monitoring system. Moreover, based on the vulnerability, the Mirai botnet can be uploaded to the camera and run. The camera becomes the Mirai botnet broiler, which can launch a large-scale DDoS attack. Because the scope of the research object is relatively small, after discovering the problems existing in the monitoring system in the campus network, we should inform the relevant departments of the school and take timely measures to protect the monitoring system. However, for large-scale protected networks, SA methods are needed to discover threat situation in time. This chapter uses DS theory to verify the feasibility and effectiveness of the proposed approach based on campus network data sources.

This chapter data source contains three kinds of data: (1) all 23 Telnet ports in the campus network in the open device and its type, IP address and other information; (2) the network camera with the weak password vulnerability of 23 Telnet in the campus network; (3) the camera which can upload Mirai virus and run it successfully through weak password vulnerability.

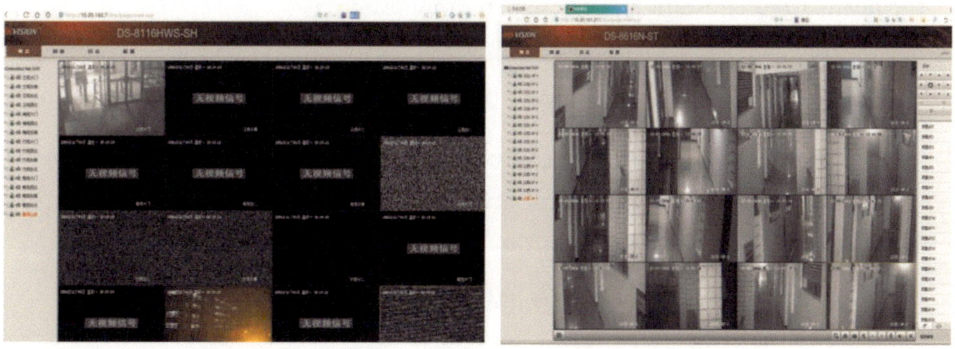

Figure 2.
Schematic diagram of campus monitoring system through weak password vulnerability.

```
10.[redacted]. 2[redacted]4 Ports: 23/open/tcp//telnet///
10.[redacted]. 2[redacted]5 Ports: 23/open/tcp//telnet///
10.[redacted]. 2[redacted]6 Ports: 23/open/tcp//telnet///
10.[redacted]. 2[redacted]7 Ports: 23/open/tcp//telnet///
10.[redacted]. 2[redacted]8 Ports: 23/open/tcp//telnet///
10.[redacted]. 2[redacted]9 Ports: 23/open/tcp//telnet///
10.[redacted]. 2[redacted]0 Ports: 23/open/tcp//telnet///
10.[redacted]. 1[redacted]6 Ports: 23/open/tcp//telnet///
10.[redacted]. 1[redacted]1 Ports: 23/open/tcp//telnet///
10.[redacted]0. [redacted]101 Ports: 23/open/tcp//telnet///
10.[redacted]0. [redacted]102 Ports: 23/open/tcp//telnet///
                ......
```

Figure 3.
Scanned device records opened on port 23.

First, scan all IOT devices opened on port 23 open and the scan results are shown in **Figure 3**. A total of 464 data opened on port 23 were recorded, including 242 camera devices. So in evidence 1, the initial trust value $m_1(V_1)$ is 242/464 $\approx$ 0.52 and $m_1(S_1)$ is 1−0.52 = 0.48.

Secondly, Scan camera equipment in school for leak detection, as shown in **Figure 4**. Among them, there are 142 camera devices with weak password vulnerabilities. So in evidence 2, the initial trust value $m_2(V_2)$ is 142/242 $\approx$ 0.59 while $m_2(S_2)$ is 1−0.59 = 0.41.

Finally, we uploaded the virus to the cameras with a weak password, and 86 camera records were uploaded successfully, as shown in **Figure 5**. So in evidence 3, the initial trust value $m_3(V_3)$ is 86/142 $\approx$ 0.61 and $m_3(S_3)$ is 1−0.61 = 0.39.

Then, the three evidences are fused by Dempster formula. If the evidence provided by the sensor scan is B, C, and D respectively, the proposition that the investigated camera in the campus network has a network security threat is called V, and the proposition that the investigated camera in the campus network is secure is called S. Then three sets of evidence are combined to calculate the confidence of proposition V as follows:

the normalized constant k is calculated as follows:

$$K = \sum_{B\cap C\cap D \neq \Phi} m_1(B)\bullet m_2(C)\bullet m_3(D)$$

$$= 0.52^*0.59^*0.61 + 0.48^*0.41^*0.39.$$

$$\approx 0.26.$$

to calculate $m(V)$ by composite formula:

$$m_1 \oplus m_2 \oplus m_3\{m(V)\}$$

$$= \frac{1}{k} \sum_{B\cap C\cap D = \{m(V)\}} m_1(B)\bullet m_2(C)\bullet m_3(D)$$

$$= \frac{1}{0.26}(0.52 * 0.59 * 0.61).$$

$$\approx 0.71.$$

to calculate $m(S)$ by composite formula:

$$m_1 \oplus m_2 \oplus m_3\{m(S)\}$$

$$= \frac{1}{k} \sum_{B\cap C\cap D = \{m(S)\}} m_1(B)\bullet m_2(C)\bullet m_3(D)$$

$$= \frac{1}{0.26}(0.48 * 0.41 * 0.39).$$

$$\approx 0.29.$$

Based on the above calculations, the ultimate trust of $m(V)$ is 0.71 and that of $m(S)$ is 0.29. Because the experimental data source in this chapter contains only campus network camera and no other devices, there is no need to estimate the threshold. In the experiment, $m(V) > m(S)$, it shows that there are serious security threats in the monitoring system of campus network by calculating the method, and the method is effective.

The prototype system based on this method is shown in **Figure 6**. The system includes a scanning module, data query, weak password management and

```
ACCOUNT FOUND: [telnet] Host: 10.[redacted]233 User: admin Password: admin [SUCCESS]
ACCOUNT FOUND: [telnet] Host: 10.[redacted] User: admin Password: admin1234 [SUCCESS]
ACCOUNT FOUND: [telnet] Host: 10.[redacted]121 User: admin Password: 12345 [SUCCESS]
ACCOUNT FOUND: [telnet] Host: 10.[redacted]122 User: admin Password: 12345 [SUCCESS]
ACCOUNT FOUND: [telnet] Host: 10.[redacted]123 User: admin Password: 12345 [SUCCESS]
ACCOUNT FOUND: [telnet] Host: 10.[redacted]124 User: admin Password: 12345 [SUCCESS]
ACCOUNT FOUND: [telnet] Host: 10.[redacted]125 User: admin Password: 12345 [SUCCESS]
ACCOUNT FOUND: [telnet] Host: 10.[redacted]63 User: root Password: 123456 [SUCCESS]
ACCOUNT FOUND: [telnet] Host: 10.[redacted]128 User: admin Password: 12345 [SUCCESS]
ACCOUNT FOUND: [telnet] Host: 10.[redacted]129 User: admin Password: 12345 [SUCCESS]
ACCOUNT FOUND: [telnet] Host: 10.[redacted]130 User: admin Password: 12345 [SUCCESS]
ACCOUNT FOUND: [telnet] Host: 10.[redacted]131 User: admin Password: 12345 [SUCCESS]
ACCOUNT FOUND: [telnet] Host: 10.[redacted]132 User: admin Password: 12345 [SUCCESS]
ACCOUNT FOUND: [telnet] Host: 10.[redacted]133 User: admin Password: 12345 [SUCCESS]
ACCOUNT FOUND: [telnet] Host: 10.[redacted]134 User: admin Password: 12345 [SUCCESS]
ACCOUNT FOUND: [telnet] Host: 10.[redacted]135 User: admin Password: 12345 [SUCCESS]
ACCOUNT FOUND: [telnet] Host: 10.[redacted]136 User: admin Password: 12345 [SUCCESS]
ACCOUNT FOUND: [telnet] Host: 10.[redacted]138 User: admin Password: 12345 [SUCCESS]
ACCOUNT FOUND: [telnet] Host: 10.[redacted]139 User: admin Password: 12345 [SUCCESS]
ACCOUNT FOUND: [telnet] Host: 10.[redacted]140 User: admin Password: 12345 [SUCCESS]
ACCOUNT FOUND: [telnet] Host: 10.[redacted]137 User: admin Password: 12345 [SUCCESS]
ACCOUNT FOUND: [telnet] Host: 10.[redacted]141 User: admin Password: 12345 [SUCCESS]
ACCOUNT FOUND: [telnet] Host: 10.[redacted]142 User: admin Password: 12345 [SUCCESS]
...
```

Figure 4.
Scanned records of cameras for leak detection in school.

```
10.[redacted]21 admin/12345 教[redacted]右侧
10.[redacted]25 admin/12345 教[redacted]
10.[redacted]34 admin/12345 教[redacted]右侧
10.[redacted]35 admin/12345 教[redacted]
10.[redacted]01 admin/admin123[redacted]图[redacted]大门
10.[redacted]03 admin/admin123[redacted]外[redacted]书阅览室（二楼）
10.[redacted]15 admin/admin123[redacted]术图书阅览室（五楼）
...
```

Figure 5.
The virus uploading records on the cameras with a weak password.

Figure 6.
Schematic diagram of network security SA prototype system based on DS evidence chain.

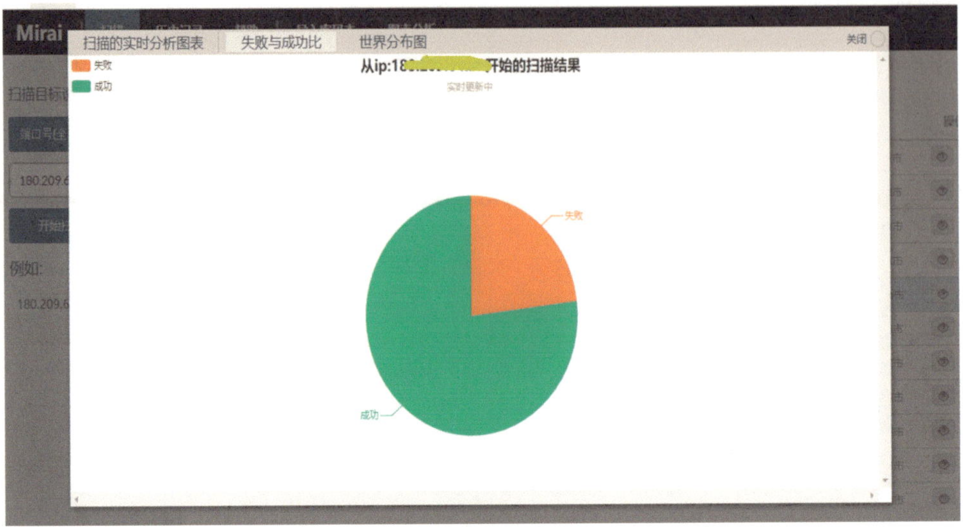

Figure 7.
Schematic diagram for the scanned result page by SA prototype system.

visualization (chart analysis) module, as shown in **Figure 7**. The scanning module integrates the automatic scanning function, as long as we input the network segment to be scanned and click "Start Scanning", the scanning can be done automatically. The buttons under the "Operation" column on the right enable you to manually access the device. For example, if the device has a weak password vulnerability, you can start shell through the "Operation" button to automatically use the weak password to login to the device for easy viewing. The "Operation" also includes manual uploading of the Mirai zombie program, etc. Data Query is designed for your viewing history scanning records; Weak Password Management for adding or removing the collected camera factory default password; Visual Analysis Module for displaying the network situation by means of geographic information, data statistics and chart, etc.

The prototype system is shown in **Figure 6**.

Figure 8.
Campus network security situation diagram based on geographic information.

DOI: http://dx.doi.org/10.5772/intechopen.89564

The security situation of campus network based on the security threat analysis of campus network camera is shown in **Figure 8**. The situation map is based on geographical location information, and the red point indicates that there is a security threat in the corresponding location of the map, which will make the administrator reminded.

6. Conclusion

This chapter first introduces the related work of SA technology, the concept, definition and formula of DS evidence theory, and then aims at the problem of slow response of network security SA to burst vulnerabilities in the network. A method of network security SA based on DS evidence theory is proposed. Finally, according to the experiment of Mirai botnet, a surveillance camera in NJUPT's campus network, it is proved that the SA method based on DS evidence theory is feasible and effective, and this method can detect the major threat in a protected network in time.

Acknowledgements

Thanks to Shu Wang, Qing Ye, Na Wang, Haichang Yao, Kui Li, Ruchuan Wang and Lu Yao for their contributions. This work is supported by the National Key Research and Development Program of China (2017YFB1401301, 2017YFB1401302, 2017YFB0202204), the National Natural Science Foundation Program of China (61373017), the Key Research and Development Program of Jiangsu Province (BE2017166), the Natural Science Foundation Outstanding Youth Fund of Jiangsu Province (BK20170100), the Open Fund of Jiangsu High Technology Research Key Laboratory for Wireless Sensor Networks (WSNLBZY201514), the 1311 Project of Nanjing University of Posts and Telecommunications.

Conflict of interest

The authors declare no conflict of interest.

Author details

Liu Shangdong[1,3,4] and Ji Yimu[1,2,3,4,5]*

1 School of Computer Science, Nanjing University of Posts and Telecommunications, Nanjing, China

2 Jiangsu High Technology Research Key Laboratory for Wireless Sensor Networks, Nanjing, Jiangsu, China

3 Institute of High-Performance Computing and Bigdata, Nanjing University of Posts and Telecommunications, Nanjing, Jiangsu, China

4 Nanjing Center of HPC China, Nanjing, Jiangsu, China

5 Jiangsu HPC and Intelligent Processing Engineer Research Center, Nanjing, Jiangsu, China

*Address all correspondence to: jiym@njupt.edu.cn

References

[1] Wu D, Cui X, Liu Q, Zhang F. Research on ubiquitous botnet. Netinfo Security. 2018;**2018**(07):16-28

[2] Xi R, Yun X, Jing S, Jin S, Zhang Y. Research survey of network security situation awareness. Journal of Computer Applications. 2012;**32**(1):1-4

[3] Gong Z, Zhuo Y. Research on network situation awareness. Journal of Software. 2010;**21**(7):1605-1619

[4] Zhang S, Liu X, Sun X. Hierarchical awareness of network security situation based on multi-source fusion. Computer Technology and Development. 2016; **26**(10):77-82

[5] Kokkonen T. Architecture for the cyber security situational awareness system. International Conference on Next Generation Wired/Wireless Networking. St. Petersburg, Russia: Springer; 2016. pp. 294-302

[6] Eiseler V, Koch R, Rodosek GD. System complexity meets decision makers: A framework for level-appropriate information processing. In: Bryant AR, Lopez J, Mills RF, editors. Proceedings of the 12th International Conference on Cyber Warfare and Security. 2017. pp. 427-431

[7] Yang Y-L. Research on network security situation awareness system based on machine learning. In: Zeng Z, Bai X, editors. Proceedings of the 2016 2nd Workshop on Advanced Research and Technology in Industry Applications. 2016. pp. 122-125

[8] Khalid HM, Peng JCH. A Bayesian algorithm to enhance the resilience of WAMS applications against cyber attacks. IEEE Transactions on Smart Grid. 2016;7(4):2026-2037

[9] Hu J, Li Z, Yao D, Yu J. Measuring botnet size by using URL and collaborative MailServers. In: Fifth International Conference on Networking and Services. 2009. pp. 161-164

[10] Evesti A, Kanstren T, Frantti T, et al. Cybersecurity Situational Awareness Taxonomy. IEEE International Conference On Cyber Situational Awareness, Data Analytics And Assessment (Cyber SA). London, UK. 2017

[11] Thomas C, Balakrishnan N. Modified evidence theory for performance enhancement of intrusion detection systems. In: Proc. International Conference on Information Fusion (ICIF), Cologne

[12] Yager RR, Fedrizzi M, Kacprzyk J, editors. Advances in the Dempster-Shafer Theory of Evidence. New York, NY: John Wiley and Sons; 1994

[13] Sabata B, Ornes C. Multisource Evidence Fusion for Cyber-Situation Assessment. 2006. pp. 624201-624201

Section 4

Blockchain

Chapter 5

Deploying Blockchain Technology in the Supply Chain

Jian Zhang

Abstract

In the rapidly evolving environment of the international supply chain, the traditional network of manufacturers and suppliers has grown into a vast ecosystem made of various products that move through multiple parties and require cooperation among stakeholders. Additionally, the demand for improved product visibility and source-to-store traceability has never been higher. However, traditional data sharing procedures in today's supply chain are inefficient, costly, and unadaptable as compared to new and innovative technology. Blockchain technology has shown promising results for improving supply chain networks in recent applications and has already impacted our society and lifestyle by reshaping many business and industry processes. In an effort to understand the integration of blockchain technology in the supply chain, this paper systematically summarizes its current status, key characteristics, potential challenges, and pilot applications.

Keywords: supply chain, blockchain, smart contract, traceability, security, digitalization

1. Introduction

The supply chain plays a crucial role in modern businesses by allowing them to achieve efficiency, responsiveness, and success. Over the past several decades, the scale of businesses has expanded, the number of geographic locales involved in the production process has grown, and product portfolios have diversified. As a result, the supply chain has grown from a traditional network of manufacturers and suppliers, to a vast ecosystem made of various products that move through multiple parties and require cooperation among stakeholders [1]. Additionally, due to the rapid evolution of e-commerce, the demand for improved product visibility and source-to-store traceability has never been higher. However, the inefficiency of data sharing in current supply chain networks has dramatically impacted the operations of retailers and manufacturers. For example, information gaps between data collected by factories and by retailers make it challenging to trace product history and offer customized products.

To overcome these challenges and improve supply chain performance, industries have explored innovative technologies that support efficient collaboration and coordination within and among different organizations [2, 3]. Among these technologies, blockchain provides a promising future and allows the supply chain to provide better visibility, transparency, and acuity of transactions throughout the entire process [4]. The blockchain technology that powers cryptocurrency has caught the attention of businesses, especially those in supply chain management. A 2017 study indicated that nearly 62% of supply chain executives claimed to have engaged with blockchain technology [5]. Although blockchain-based applications in the supply chain are still in their early stages, we believe this technology will significantly remodel the supply chain system [6–8]. Analysts forecast that blockchain technology can help supply chain management gain one-third improvement in most of its common processes [9]. A blockchain network is as a distributed ledger—transactions are contained in blocks that are linked together in chronological order to form a tamper-proof chain, which is usually stored in all network nodes [10, 11]. As such, blockchain technology provides a means to create tamper-proof logs of business activities and transactions [12]. Transaction data are immutable because they cannot be tampered with once they are distributed, accepted, and validated by network consensus and stored in the blocks [13]. By eliminating intermediaries to achieve trust among all stakeholders, efficiency improves and cost is reduced for the entire supply chain.

Despite the general acceptance that blockchain technology facilitates faster, more easily auditable interactions and allows for the exchange of immutable data among supply chain partners [14], it will take time for this technology to be adopted and to revolutionize the supply chain. Currently, most applications of blockchain are conceptual expositions, and empirical evidence on the implementation of it is limited [15]. Furthermore, few studies have been conducted on the challenges of deploying blockchain in the supply chain, such as organizational readiness, technical expertise, scalability, and compatibility with existing systems. Therefore, this study will provide a systematic analysis of how blockchain technology fits in the supply chain network and discuss potential challenges with its implementation.

2. Supply chain

2.1 Overview

Supply chain encompasses the end-to-end flow, including the physical and correlated data flow of raw material, products, information, and money. It plays a unique and critical role in businesses and determines the performance of organizations. Supply chain manages or is involved in sourcing, procurement, manufacturing, distribution, and logistics, and, thus, affects speed-to-market, the cost of a product, service perception, and capital requirements in businesses [16]. Supply chain integrates a set of fragmented and often geographically discrete processes into a cohesive system to deliver value to the customer. The core functions and operations of a typical supply chain network are illustrated in **Figure 1**.

2.2 Problems with today's supply chain

Evolving customer requirements, challenges from competition, geographically separated operations, and the adoption of new business models (such as

Figure 1.
Supply chain and operations.

e-commerce) make the current supply chain a highly complex system. Over the past decade, e-commerce and hand-held digital devices have substantially changed the daily lives of people, especially in the ways they shop. There is an ever-increasing demand for customized products, a simplified and efficient shopping experience, and transparency about the value and provenance of goods. These needs bring new opportunities to businesses but impose significant challenges to current supply chains. These outdated supply chains struggle to improve demand management, to provide data visibility for the entire flow, or to track goods from raw material to end consumer—all of which are tremendously complex. Furthermore, the old technology of today's supply chain fails to provide adequate risk management, to reduce costs, or to meet rapidly changing market requirements. We summarize the main challenges in current supply chains here:

Lack of traceability: In the last few years, traceability has become crucial for supply chains to address, especially in regard to customer service and planning and forecasting in business operations. However, it is difficult to deploy a centralized system in an interconnected network, especially where trust among participants is limited. Instead, there are several discrete systems among involved parties that consist of various databases that impede product tracking throughout the entire supply chain network [17].

Stakeholder distrust: Trust is an essential factor in supply chain management, and an effective supply chain network must be built on a solid foundation of it [18]. However, distrust among participants is the single greatest obstacle to improving supply chain networks [19]. Consequently, most stakeholders in the network primarily rely on third-party intermediaries to serve as agents of trust and to verify transactions, which dramatically increase operational cost and reduce process efficiency.

Limited transparency: The term "transparency" in the supply chain refers to the extent to which all stakeholders own a shared understanding of and access to accurate and adequate information about products [20, 21]. A transparent supply chain network improves trust among stakeholders and guarantees the integrity of products and associated data. However, the discrete databases in current supply chain networks offer minimal transparency, and most of the useful information in them is lost when products and data are transferred from one stakeholder to another. Furthermore, there are issues with inconsistent data sharing, relying on paper documentation, and inadequate interoperability. These critical challenges remain despite years of significant research investment. The crisis of Chipotle Mexican Grill outlets [7] is an important and sad example of how the current supply chain system is inefficient at, and possibly incapable of, offering transparency throughout the entire lifecycle of products.

Outdated means of data sharing: In current supply chain networks, data are shared between many organizations using paper-based documentation. Oftentimes, important documents, such as bills of lading, letters of credit, invoices, insurance policies, and various certificates, must travel with their associated goods around the world [22]. For example, about 200 communications were needed for Maersk, a global transport and logistics company, to complete a single shipment of frozen goods from Mombasa to Europe in 2014 [23]. These communications created a stack of documents about 25 centimeters in height [24]. Constrained by this outdated and inefficient data sharing method, ships and airplanes are often delayed in ports when the paperwork does not match the carried goods [22].

Compliance challenges: Currently, businesses have to meet increasingly strict regulatory standards to provide safe products and services to customers. Recently, the U.S. Food and Drug Administration and Federal Trade Commission adopted several standards to increase food safety and offer full visibility of food flows in the supply chain. However, under current supply chain processes, it is difficult to obtain this information from a variety of stakeholders and to develop a database that complies with new standards.

3. Blockchain technology

Blockchain is an innovational technology that enhances customer service, drives end-to-end value, and increases the efficiency of operations [25]. Additionally, it allows distrusting or unfamiliar stakeholders to create shared and secure data records [26]. In sum, when an exchange of valuable data and goods is necessary, blockchain technology expedites transactions, streamlines the process, enhances transparency, reduces waste, and, ultimately, reduces cost [27]. Consequently, new types of internet and associated business models have been built off of this robust technology [22]. Blockchain promises to be the primary driver of secure and efficient economic and social systems in the future.

3.1 What is blockchain technology?

3.1.1 Chained architecture

The basic concepts of blockchain were introduced by Satoshi Nakamoto in Bitcoin [28], a digital cryptocurrency that can work without the need of a trusted intermediary. It offers a distributed ledger that tracks and sustains a tamper-proof record of transactions in a decentralized network. In essence, it is a unique database system that is created, replicated, synchronized, and maintained by all participants in the decentralized network. Blockchain operates in a decentralized peer-to-peer network [29] to validate and store all transactions in a consensus that is agreed upon by all nodes in the network, without any central authority to validate the transaction (as with an intermediary). All completed and validated transactions are logged in the distributed ledger in a verifiable, secure, transparent, and permanent manner along with a timestamp and other details [30]. In this way, the exchange of tangible and intangible data and assets among participants can be recorded digitally. Each stakeholder maintains a copy of the synchronized ledger, which prevents a single point of system failure or data loss [22]. When changes are made, such as adding a new block, all copies in the network are simultaneously updated, and records are permanently registered in all ledgers [31]. These changes are stored into blocks that create a chain [32], where a block is linked to the preceding one by storing its hash (a unique data that is mapped from the given block) [33]. **Figure 2** shows the fundamental chained architecture of a blockchain network.

In **Figure 2**, notice that except for the first block (called the genesis block), each block has its hash as a unique ID that includes the hash of the previous block. In this way, a chronological chain is formed. Additionally, the hash mechanism provides enhanced data security. Usually, a block stores a set of time-stamped transactions that are validated by stakeholders in the network. Once it gains consensus, the block is accepted and stored by all parties in the blockchain and can no longer be modified. Therefore, trust in and transparency of transactions between organizations are significantly improved.

3.1.2 Permissionless vs. permissioned blockchain

Since the introduction and success of Bitcoin, many blockchain-based platforms can be categorized as either a permissionless or permissioned blockchain. Virtually,

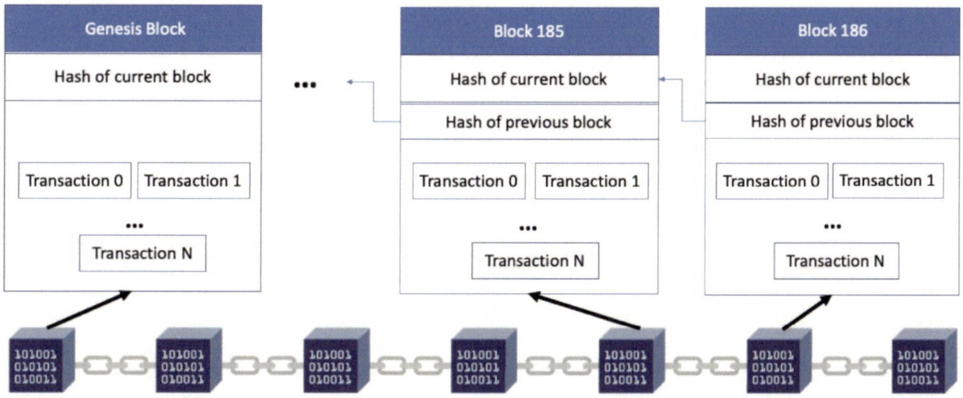

Figure 2.
The architecture of a data chain in a blockchain network.

anyone can join and participate anonymously in a permissionless blockchain network. Accordingly, it is also called a public blockchain, and these two notions will be used interchangeably in the remaining sections. Within this type of network, trust among users is limited or nonexistent. To overcome this lack, miners (detailed later) are introduced to validate transactions.

In contrast, permissioned blockchain is a network for a group of identified users operating under a governance model, called a consensus, to improve transactional trust. To join this type of network, new users need permission from the majority of the group or a delegated user; hence, it is also called a private blockchain, and we use both notions interchangeably in this paper. These networks facilitate trust among users and do not require costly miners. More efficient consensus protocols (such as the Byzantine fault tolerant protocol) validate data, improve network throughput, and reduce the latency of transactions.

3.1.3 Key characteristics

Blockchain technology has many unique features that allow for the creation of a verifiable, secure, transparent, and immutable distributed ledger, the core characteristics of which are summarized as follows:

1. **Versatile value exchange**: Blockchain provides a secure and efficient platform for recording the transactions of intellectual property rights, the provenance of services and goods, asset ownership, cryptocurrency exchange, and more.

2. **Distributed governance**: A blockchain network is not controlled by any designated authority, organization, or person, and the need for trusted intermediaries to verify transactions is eliminated. It is a distributed database that provides secure and validated data for all participants in the network simultaneously. Thus, there is full transparency along the entire stream of transactions, and assets and data can be transferred between several organizations in a quick and efficient way.

3. **Decentralized architecture**: The ledger is decentralized and stored in all nodes (i.e., individual stakeholder databases) of the network, and failure of it at a central infrastructural point is not possible. Therefore, it fosters a robust network that improves the quality, reliability, and availability of services and information.

4. **Logically centralized**: With only one transaction record shared with and agreed upon by all participants, a blockchain network behaves like a logically centralized system.

5. **Data transparency**: Blockchain technology allows for a highly transparent network that is visible to each stakeholder at all times. This dramatically reduces the chances of illegal transactions.

6. **Immutable data**: Once a block with a set of transactions is verified by the consensus and stored in the chain, the encapsulated data can no longer be modified.

7. **Enhanced data security**: Blockchain technology utilizes asymmetric cryptography and digital signature algorithms to ensure data security and individual identity.

3.1.4 Main components and data flow

To cater to the vastly different needs of unique businesses and users, many blockchain networks are created, and each contains a slightly different set of features; however, a basic foundation remains the same for all. As an example, we use Bitcoin, the first and the most successful permissionless blockchain system, to illustrate the key components of typical data flow in a blockchain network:

Block: A data structure that is used to collect a set of transactions and is protected by adding a hash value to ensure the integrity of stored data. It is an essential component and is deployed in all blockchain networks.

Digital wallet: A secure repository for a user to store the private and public key pair. It interacts with the Bitcoin network so a user can receive and send digital currency (Bitcoins) and monitor their balance.

Node: A client who participates in transactional activities on the blockchain network. First and most importantly, a node owns a complete and permanent copy of the ledger that consists of all historical transactions. It works as a cornerstone to store a full copy of the tamper-proof ledger in each node in a blockchain network. Second, a node contributes to the network by broadcasting transactions and enabling miners to validate and create blocks.

Miner: A miner, a special user in the Bitcoin network, collects and validates all broadcasted transactions and creates new blocks. It competes with other miners in the network to solve a mathematical puzzle, widely known as a proof-of-work problem. The first to win the puzzle adds a new block to the chain and gains a specific amount of reward, such as a small number of Bitcoins. When a block is added, all nodes synchronize their local copy, ensuring their ledger is up-to-date. A miner or mining procedure is used for validation in many permissionless blockchains, whereas validation is executed by nodes under the control of a consensus in most permissioned blockchains.

Consensus: An agreement between nodes in a blockchain network that submits transactional information, and is one of the most critical components of blockchain technology. A blockchain network is updated via the deployed consensus protocol to ensure that transactions and blocks are ordered correctly, to guarantee the integrity and consistency of the distributed ledger, and, ultimately, to enhance trust between stakeholders (nodes). Additionally, a consensus algorithm can help a distributed or decentralized network unanimously make a decision [11, 29]. Prevalent consensus algorithms include proof-of-work, proof-of-stake, Byzantine fault tolerance, delegated proof-of-stake, proof-of-elapsed time, and proof-of-authority matched [34, 35].

In a typical open and permissionless blockchain network such as Bitcoin, when a user starts a transaction, the digital wallet verifies and signatures the transaction before broadcasting it to all nodes in the network. The verified transaction is added to a block that collects a set of new transactions. Miners validate the block, and once validated, the block is added to the existing blockchain by all nodes. This completes the transaction. The following is an illustration of typical data flow within the Bitcoin network:

A typical permissioned blockchain follows a similar data flow to that illustrated in **Figure 3**, where a signature is added to the transaction, which is then submitted or broadcasted to the network and added to a block. After the block is validated, the transaction is permanently stored in the chain. Permissioned blockchain differs from permissionless blockchain by how blocks and transactions are validated. To gain better performance and lower latency, most permissioned blockchain networks deploy efficient consensus protocols (e.g., the Byzantine fault tolerance consensus used by Hyperledger Fabric) that nodes use for validation.

3.2 What is a "smart contract"?

The term "smart contract" was first proposed by Nick Szabo, and defined as "a set of promises, specified in the digital form, including protocols within which the parties perform on these promises" [36]. The smart contract concept was integrated into Ethereum's blockchain network to facilitate, verify, and enforce contract negotiations and to improve the contract performance. Before transactions are conducted in a blockchain network, a smart contract that defines the conditions,

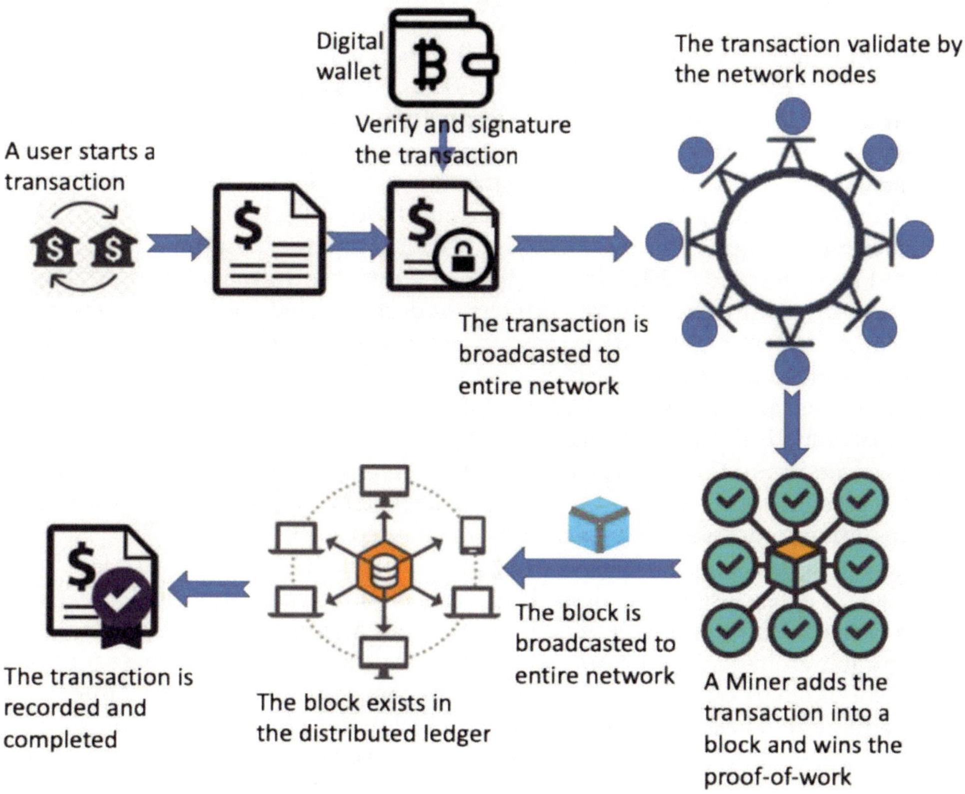

Figure 3.
Data flow of an open blockchain network [27].

obligations, rights, and concepts between stakeholders is created. This information is recorded as executable computer code to reduce ambiguity. Smart contracts are stored and shared in a distributed ledger that all participants have access to. These contracts automatically self-execute when all of the pre-set conditions are satisfied within a blockchain network. Thus, stakeholders who agreed upon a smart contract have more trust for each other and have a reduced risk of error and fraud [37]. The following details additional advantages of smart contracts:

- **Cost-saving**: by eliminating intermediaries and reducing process time;
- **Accurate**: all agreements, conditions, etc. are recorded in terms of computer codes that provide a more accurate and efficient means of information storage;
- **Speedy**: Whenever the pre-defined conditions are met, the smart contract is executed autonomously and in real-time;
- **Transparent**: Smart contracts are available and fully visible to all participants involved in the network; and
- **Secure**: Smart contracts are stored using encryption and are distributed on all nodes of the blockchain network simultaneously.

3.3 Existing blockchain platforms and applications

There are many blockchain platforms with different consensus algorithms, development tools, and programming languages [38]. We introduce a few important blockchain platforms and applications herein.

Bitcoin: The initial and most famous blockchain network to offer crypto-currency transactions. It was launched in 2009 and has rapidly grown to be a significant currency system both online and offline. Since the mid-2010s, increasingly more businesses have begun accepting Bitcoin as payment. At the time of this writing (March 2019), the market capitalization of Bitcoin was about $68 billion [39]—it takes around 10 minutes to create a new block [40].

Ethereum: An open-source blockchain platform that was introduced by Buterin [41] and first launched in 2015. It is the first, and possibly the most advanced, blockchain network to introduce smart contracts for decentralized applications (Dapps). The primary Ethereum network serves as a public blockchain network; however, it is also possible to create a private blockchain network based on Ethereum. Quorum [42] is one such example and deploys the Ethereum network to create an enterprise-ready distributed ledger and smart contract platform, both of which contribute to faster processing. In Ethereum's main network where a majority of transactions take place, it takes about 10–15 seconds to create a new block [43]. However, the number of transactions processed per minute is still as limited as Bitcoin.

Hyperledger fabric: An open-source, private blockchain network that is designed for enterprise applications. Hyperledger Fabric was established under the Linux Foundation and is maintained by a variety of organizations [44]. It employs a configurable architecture that provides various features, such as distributed ledger frameworks, smart contract engines, pluggable consensus protocols, user interfaces, and more. These versatile characteristics allow for a broad range of business applications, including finance, insurance, supply chain, healthcare, and human resources.

Skuchain: A blockchain network that is designed for enterprise supply chains in global trade [45]. It creates a zero-knowledge collaborative platform for global supply chains and provides precise control in inventory procurement across all partners, reducing friction and the costs of supply chain processes.

Sweetbridge: A blockchain-based application that enables real-time financial systems to assure transactional data are trustworthy between different parties. It integrates trusted identity, smart legal contracts, smart accounting, and payment rails into a transaction for all parties to see in real-time [46].

Zervnetwork: A decentralized trading platform based on blockchain technology. It aims to provide frictionless transactions among all participants within the defense industry [47].

IOTA: An open-source distributed ledger that is being built to power the future of the Internet of Things (IoT) with feeless microtransactions and data integrity for machines [48].

4. Chain integration

In recent years, Blockchain technology has been recognized as a critical technology with inherent capabilities to dramatically improve supply chain efficiency [49–51]. A study from Eye for transport stated that more than 16% of the 300 companies surveyed agree that data interchange, tracking, and visibility are the foremost reasons to deploy blockchain technology in the supply chain [52]. However, we discuss the benefits, challenges, and risks of integrating blockchain technology in the supply chain and introduce several pilot initiatives below.

4.1 Benefits to supply chain

The adaptation of blockchain technology can significantly alleviate or even eliminate the aforementioned problems in today's supply chain. Blockchain technology empowers the supply chain with improved efficacy, efficiency, and transparency and reduced transactional time and cost. There are many ways blockchain technology benefits the supply chain:

Advanced traceability: With the adoption of blockchain technology, traceability within the supply chain is greatly improved; it produces a fully auditable trail of all items flowing through the network. Combined with IoT-based devices, such as RFID technology, a blockchain-enabled supply chain can automatically collect the item-level data of massive quantities of products in real-time. Additionally, this information is associated with timestamps and collection locations to form an audit trail that is complete, accurate, and easy-to-access, from the product's origin to the customer. Furthermore, thanks to the immutability of blockchain data and the digital signatures required to confirm information ownership, data stored in this chain offers a secure and full history of any item in the entire supply chain. In the event of a compromised product, improved traceability enables the source of the issue to be identified more quickly, which reduces the cost of recalling products and improves disruption resolution between stakeholders. Advanced traceability gives stakeholders and customers more confidence in a product's authenticity and quality.

Improved transparency: Blockchain technology provides reliable identity management in the supply chain [53] by enabling all parties to know who is performing what actions, at what time, and where. This information is stored and shared in distributed ledgers that can be conveniently accessed by involved and authenticated stakeholders. Through the integration of physical and digital flows across the supply chain, the connectivity of multiple trading partners will improve [54, 55]. Therefore, a blockchain-enabled supply chain with its transparent and complete inventory of product flow helps businesses make better forecasts and decisions. Additionally, improved transparency serves as a powerful tool for fighting fraud and counterfeiting.

Boosted efficiency: One of the primary motivations for implementing blockchain technology is to replace the outdated, paper-heavy processes in place today. As one of the benefits of digitalization, the logically centralized data ledger provides up-to-date local copies to all stakeholders within the network. All transactions are committed and immediately validated by all involved parties, and data are automatically synchronized to each party's local copy. Blockchain technology makes it safer and faster to maintain the quality of transactions and associated data [56] by reducing human error and eliminating the need for third-party intermediaries and for local ledger reconciliation. Finally, the autonomous and self-executing blockchain-based smart contract replaces tedious processes and improves flexibility in supply chain management.

Greater security: It is nearly impossible to impact blockchain technology through hacking attacks like those that threaten centralized databases of intermediaries (e.g., banks). It is structured so that when there is an attempted hack into a specific block, all preceding blocks in the entire history must also be tampered with. Thus, blockchain provides a more secure way to maintain a log of business activities and transactions [12].

Enhanced trust: The transactions of a blockchain-based supply chain are created and recorded based on peer-to-peer interaction that can be trusted by the associated digital signatures. Additionally, a reliable identity management mechanism [53] allows for the collection of time, location, and other data at every action on a product in the supply chain. All data are synchronized to all stakeholders in real-time, which enhances trust among stakeholders within the supply chain network.

Easy compliance: A blockchain-enabled supply chain network records all transactions with precise details, such as timestamps, environmental conditions, and location. These accurate, tamper-proof records can serve as the source of a business's data integrity and be easily accessed for regulations and compliance.

4.2 Challenges with blockchain technology

Although blockchain technology is widely recognized as a promising solution for issues with today's supply chain, the application of it requires significant changes in both technological and cultural contexts. Additionally, more comprehensive evaluations of it are needed to unveil and address its challenges before the full potential of this new technology can be realized [22, 57].

Throughput and performance: Due to its decentralized architecture, each transaction is approved by all or a majority of nodes in a blockchain network. This approval process limits the throughput of a blockchain network; for example, Bitcoin, a public blockchain, can only process from 3 to 30 transactions per second. However, a private blockchain-based supply chain network must process far more transactions, possibly thousands per second, for the entire system. Thus, it is imperative to improve the transaction capacity of blockchain technology for full scalability. Fortunately, a private blockchain network's ability to improve the throughput of transactions may mitigate this processing challenge.

Standardization: Standardization is a critical concern for the adoption of blockchain technology in the supply chain. In essence, this technology offers a ubiquitous and general-purpose platform for digital data sharing and permanent storage. Interestingly, a major question still remains: what content and format should be adopted for transactional data that facilitates interpretation by all participants? A data standard must be established and agreed upon by the entire supply chain community. However, there is no existing standard that can be adapted for this purpose.

In recent years, much effort, such as EPCIS [58] that proposed GS1, has been made to overcome this gap, however, it is still not widely accepted and implemented in supply chains.

Data privacy: The immutability and transparency of blockchain technology raise a concern with data privacy when deployed for supply chains. Once data are stored in blockchains it cannot be changed, and, thus, it is imperative that a reliable mechanism that protects users' privacy is designed. The task of balancing an individual's right to privacy in an open blockchain network is very challenging. Currently, most blockchain networks, such as Bitcoin, provide limited control to users over the data and where they can transfer it to [22]. Most networks offer only pseudonymity to its users for privacy, so, although transactions are public for all nodes, the real identity of their owners is never revealed. This is unacceptable for supply chains, as nobody is willing to leak information to competitors about Confidential detail or the amount of merchandise moving in a network. Furthermore, with the limited number of stakeholders in the supply chain, it would be easy to figure out the owner of the transactional data and anonymity would disappear. To address this, private blockchain technology (such as Hyperledger Fabric) can support the creation of a channel for limited and trusted parties who are involved in specific transactions [44]. In this way, an unauthenticated user is forbidden to join the channel or access its data. It should be noted that a blockchain network can be designed to only serve as metadata of the workflow and the contents and details of all transactions within it are stored in external data repositories. Therefore, this technology provides a log of transactions on which no private data are stored [13].

4.3 Pilot initiatives

Since late 2016, retail giants Walmart and IBM worked together for a pilot project to develop a blockchain-based system for tracking produce in the U.S. and pork in China. The project traced each product and collected its associated data, including origin farm/factory, storage temperature, and serial number. With this technology, tracking reports for each product were produced within minutes and the speed and accuracy of identifying and recalling contaminated food products were significantly improved [59]. On May 31, 2017, Walmart released the results of this pilot project and reported that blockchain technology helped them trace the origin of Chinese pork and U.S. mangoes in 2.2 seconds, which would normally take as long as several weeks in a traditional supply chain platform [60].

Intel conducted a public demo to explore the implementation of blockchain technology for tracking seafood in the supply chain. They aimed to create a network that assists multiple parties with food storage condition (i.e., temperature) control and with tracking food from sea to table. Several public records of this project are available on the Traceability Blockchain website [61]. These records detail how to use blockchain technology to collect seafood product data (i.e., locations, timestamps, owners, temperatures, etc.) from fishermen, transports, and restaurants within the entire supply chain network. This seafood blockchain can foster more trust between customers and sellers, improve and expedite the food safety network, and enhance consumer experiences.

In 2018, el Maouchi introduced TRADE, a fully transparent and decentralized traceability system for the supply chain that leverages blockchain technology [17]. It is a single system in which multiple participants can transfer and track products flowing through the supply chain. Additionally, it enables customers and other parties in the system to view and verify product data. Experiments show that each actor on the TRADE system can create about 351 and validate 437 transactions per second.

Since August 2018, IBM and Maersk (the world's largest shipping company) have teamed up to create TradeLens, a blockchain-based system for the global supply chain. TradeLens aims to create a platform for multiple trading parties to securely share databases containing massive amounts of transactional information, and to build a more collaborative environment for global trading. This system is a powerful tool for establishing a single and consistent shared status of each transaction in near real-time while maintaining stakeholder confidentiality. Reports show that TradeLens significantly reduced delays caused by documentation errors and reduced the transit time associated with shipping packaging materials to manufactures in the U.S. up to 40% [62].

5. Conclusion

Introduced in 2009 as the foundation for Bitcoin, blockchain technology shows the significant capacity to benefit today's supply chain. It provides a decentralized platform that shares any type of transaction and that records information with an immutable and permanent historical trail. We believe it has a significant future in the supply chain, as it promises to deliver an efficient, transparent, and collaborative network for organizations to quickly and securely share data across the variety of supply chain sectors and processes. This technology allows businesses to build a more flexible and responsible supply chain, and to robustly address new external and internal challenges.

Author details

Jian Zhang
RFID Lab, Auburn University, Auburn, AL, United States of America

*Address all correspondence to: jzz0043@tigermail.auburn.edu

References

[1] Aste T, Tasca P, Di Matteo T. Blockchain technologies: The foreseeable impact on society and industry. Computer. 2017;**50**(9):18-28

[2] Farooq S, Obrien C. A technology selection framework for integrating manufacturing within a supply chain. International Journal of Production Research. 2012

[3] Williamson EA, Harrison DK, Jordan M. Information systems development within supply chain management. International Journal of Information Management. 2004

[4] Pilkington M. Blockchain technology: Principles and applications. In: Research Handbook on Digital Transformations. Elgaronline; 2016. p. 225

[5] 2017 Supply Chain Trends Recap. Eyefortransport [Online]. 2017. Available from: https://www.eft.com/content/2017-supply-chain-trends-recap

[6] Biswas K, Muthukkumarasamy V, Tan WL. Blockchain Based Wine Supply Chain Traceability System Blockchain View project Innovative Applications of Blockchain Technology View project Blockchain Based Wine Supply Chain Traceability System. 2017

[7] Kshetri N. 1 Blockchain's roles in meeting key supply chain management objectives. International Journal of Information Management. Elsevier; 2018;**39**:80-89

[8] Lu Q, Xu X. Adaptable blockchain-based systems: A case study for product traceability. IEEE Software. 2017;**34**(6):21-27

[9] Kamble S, Gunasekaran A, Arha H. Understanding the blockchain technology adoption in supply chains—Indian context. International Journal of Production Research. Taylor & Francis; 2019;**57**(7):2009-2033

[10] Underwood S. Blockchain beyond bitcoin. Communications of the ACM. 2016;**59**(11):15-17

[11] Gupta M. Blockchain for Dummies. 2nd IBM Limited ed. Hoboken, NJ, US: John Wiley & Sons, Inc.; 2018

[12] Lemieux VL. Trusting records: Is blockchain technology the answer? Records Management Journal. 2016;**26**(2):110-139

[13] Litke A, Anagnostopoulos D, Varvarigou T. Blockchains for supply chain management: Architectural elements and challenges towards a global scale deployment. Logistics. 2019;**3**(1):5

[14] IBM. The Benefits of Blockchain to Supply Chain Networks. [Online]. 2016. Available from: https://www.techrepublic.com/resource-library/whitepapers/ibm-the-benefits-of-blockchain-to-supply-chain-networks/

[15] Ying W, Jia S, Du W. Digital enablement of blockchain: Evidence from HNA group. International Journal of Information Management. 2018;**39**(2017):1-4

[16] Kehoe L, O'Connell N, Andrzejewski D, Gindner K, Dalal D. When two chains combine supply chain meets blockchain. Deloitte. 2017:2-15

[17] el Maouchi M, Ersoy O, Erkin Z. TRADE: A transparent, decentralized traceability system for the supply chain. In: Proceedings of 1st ERCIM Blockchain Workshop 2018. European Society for Socially Embedded Technologies (EUSSET). 2018

[18] Tyndall G, Gopal C, Partsch W, Kamauff J. Supercharging supply chains.

New Ways to Increase Value Through Global Operational Excellence; 1998

[19] Poirier CC. Advanced Supply Chain Management: How to Build a Sustained Competition. Berrett-Koehler; 1999

[20] Deimel M, Frentrup M, Theuvsen L. Transparency in food supply chains: Empirical results from German pig and dairy production. Journal on Chain and Network Science. 2008;**8**(1):21-32

[21] Pant RR, Prakash G, Farooquie JA. A framework for traceability and transparency in the dairy supply chain networks. Procedia-Social and Behavioral Sciences. 2015;**189**:385-394

[22] Chang Y, Iakovou E, Shi W. Blockchain in global supply chains and cross border trade: A Critical Synthesis of the State-of-the-Art, Challenges and Opportunities. 5 Jan 2019. arXiv preprint arXiv:1901.02715

[23] I.N. Release. Maersk and IBM Unveil First IndustryWide CrossBorder Supply Chain Solution on Blockchain [Online]. 2017. Available from: https://www-03.ibm.com/press/us/en/pressrelease/51712.wss#feeds

[24] Allison BI. Shipping giant Maersk tests blockchain-powered bill of lading [Online]. 2016. Available from: http://www.ibtimes.co.uk/shipping-giant-maersk-tests-blockchain-powered-bills-lading-1585929

[25] Agarwal S. Blockchain Technology in Supply Chain and Logistics. Cambridge, MA, US: Massachusetts Institute of Technology; 2018

[26] The Economist Staff. The great chain of being sure about things. The Economist. 2015. pp. 1-10

[27] Wasserman P. Santander's InnoVentures Distributed Ledger Challenge: Decoding Blockchain [Online]. 2016. Available from: http://www.sachsinsights.com/santanders-innoventures-distributed-ledger-challenge-decoding-blockchain

[28] Nakamoto S. Bitcoin: A Peer-to-Peer Electronic Cash System. [Online]. 2008. Available from: https://bitcoin.org/bitcoin.pdf

[29] Chen Y. Blockchain tokens and the potential democratization of entrepreneurship and innovation. Business Horizons. 2018;**61**(4):567-575

[30] Holotescu C. Understanding blockchain technology and how to get involved. Science Conference. 2018;**2018**:300-308

[31] Iansiti M, Lakhani KR. The truth about blockchain. Harvard Business Review. 2017;**95**(1):118-127

[32] L L, Zhi Li JH, Wang WM, Liu G. Toward open manufacturing: A cross-enterprises knowledge and services exchange framework based on blockchain and edge computing. Industrial Management & Data Systems. 2018;**118**(1):303-320

[33] Catalini C, Gans JS. Some simple economics of the blockchain. No. w22952. National Bureau of Economic Research. 2016

[34] Baliga A. Understanding Blockchain Consensus Models. Whitepaper. 2017. pp. 1-14

[35] Cachin C, Vukolić M. Blockchain consensus protocols in the wild. 6 Jul 2017. arXiv preprint arXiv:1707.01873

[36] Szabo N. Smart contracts: Building blocks for digital markets. Extropy: Journal of Transhumanist Thought. 1996;**18**(16)

[37] Chu Y, Ream J, Schatsky D. Getting smart about smart contracts. Deloitte CFO Insights. [Online]. 2016. Available from: https://www2.deloitte.com/tr/

en/pages/finance/articles/cfo-insights-getting-smart-contracts.html

[38] Body A. Blockchain: How to choose the right tech for your business [Online]. 2018. Available from: https://medium.com/@abody/blockchain-how-to-choose-the-right-tech-for-your-business-aa4597d7ee7c

[39] CoinMaketCap [Online]. Available from: https://coinmarketcap.com/

[40] Blockchain Luxembourg [Online]. Available from: https://www.blockchain.com/stats?

[41] Buterin V. A next-generation smart contract and decentralized application platform. White Paper. Jan 2014

[42] Morgan JP. Quorum [Online]. Available from: https://www.jpmorgan.com/global/Quorum

[43] Ethereum Status [Online]. Available from: https://ethstats.net/

[44] Hyperledger Fabric [Online]. Available from: https://hyperledger-fabric.readthedocs.io/en/latest/whatis.html

[45] skuchain [Online]. Available from: http://www.skuchain.com

[46] sweetbridge [Online]. Available from: https://sweetbridge.com.

[47] zervnetwork [Online]. Available from: https://zervnetwork.com/

[48] iota [Online]. Available from: https://www.iota.org/

[49] Kim HM, Laskowski M. Toward an ontology-driven blockchain design for supply-chain provenance. Intelligent Systems in Accounting, Finance and Management. 2018;**57**(1):18-27

[50] Tian F. An Agri-food supply chain traceability system for China based on RFID & blockchain technology. In: 2016 13th International Conference on Service Systems and Service Management, ICSSSM 2016. IEEE; 2016. pp. 1-6

[51] Abeyratne SA, Monfared RP. Blockchain ready manufacturing supply chain using distributed ledger. International Journal of Research in Engineering and Technology. 2016

[52] Outlier Ventures Blockchain-Enabled Convergence: Understanding the Web 3.0 Economy [Online]. 2016. Available from: https://gallery.mailchimp.com/65ae955d98e06dbd6fc737bf7/files/Blockchain_Enabled_Convergence.01.pdf

[53] Alam M. Why the auto industry should embrace Blockchain [Online]. 2016. Available from: http://www.connectedcar-news.com/news/2016/dec/09/why-auto-industry-should-embrace-blockchain/

[54] Takahashi BR. How can creative industries benefit from blockchain? Mckinsey [Online]. 2017. Available rom: https://www.mckinsey.com/industries/media-and-entertainment/our-insights/how-can-creative-industries-benefit-from-blockchain

[55] Min H. Blockchain technology for enhancing supply chain resilience. Business Horizons. 2019;**62**(1):35-45

[56] Verhoeven P, Sinn F, Herden T. Examples from blockchain implementations in logistics and supply chain management: Exploring the mindful use of a new technology. Logistics. 2018;**2**(3):20

[57] Thurner T. Business innovation through blockchain: The B3 perspective. Foresight. 2018;**20**(5):583-584

[58] GS1. EPCIS [Online]. May 2014. Available from: https://www.gs1.org/standards/epcis

[59] Yiannas F. A new era of food transparency with Wal-Mart center in

China. International Journal Food of Safety News [Online]. 2017. Available from: https://www.foodsafetynews.com/2017/03/a-new-era-of-food-transparency-with-wal-mart-center-in-china/

[60] Nation J. Walmart tests food safety with blockchain traceability. ETHnews [Online]. 2017. Available from: https://www.ethnews.com/walmart-tests-food-safety-with-blockchain-traceability

[61] Traceability Blockchain Website [Online]. 2017. Available from: https://provenance.sawtooth.me/#

[62] IBM Corporation. Maersk and IBM introduce TradeLens blockchain shipping solution. IBM Newsroom [Online]. 2018. Available from: https://newsroom.ibm.com/2018-08-09-Maersk-and-IBM-Introduce-TradeLens-Blockchain-Shipping-Solution

Chapter 6

Blockchain Applications in Cybersecurity

Oscar Lage, Santiago de Diego, Borja Urkizu, Eneko Gómez and Iván Gutiérrez

Abstract

Blockchain has been widely known thanks to Bitcoin and the cryptocurrencies. In this chapter, we analyze different aspects that relate to the application of blockchain with techniques commonly used in the field of cybersecurity. Beginning by introducing the use of blockchain technology as a secure infrastructure, the document delves into how blockchain can be useful to achieve several security requirements, common to most applications. The document has been focused on some specific cybersecurity disciplines to maintain simplicity: backup and recovery, threat intelligence and content delivery networks. As illustrated, some projects and initiatives are in the process of joining these two fields to provide solutions to existing problems.

Keywords: blockchain, DLT, trust, cybersecurity, IoT, IIoT

1. Introduction

Blockchain is a very-known term, which was used for the first time in [1], where Satoshi Nakamoto described Bitcoin in 2008. Bitcoin is the best-known implementation of blockchain, and it is basically the implementation of a cryptocurrency. However, blockchain is much more than that, being seen as the service and structure behind cryptocurrencies to maintain records for currency transactions between untrusted participants. Nowadays, in addition to cryptocurrencies (hundreds of currencies exist today that use blockchain technology or derivatives), many other application areas rely on blockchain technology like energy trading, health, supply chain, manufacturing, identity management, e-government, etc.

Blockchain presents itself as a distributed ledger, referring this concept to the way a database is shared between several participants on a peer-to-peer network, without a central authority overseeing the process. In the case of blockchain, this ledger is arranged, as its name suggests, in an ordered chain of blocks, each of which agglutinates transactions in order. A block, therefore, is basically a structure composed of a header and a body containing transactions in order. Blocks are timestamped and signed by its creator. The way these blocks constitute a chain is through a pointer to the previous block; the header of each block contains a cryptographic hash of the previous block so that a block is linked to the previous one (while ensuring the immutability of that previous block). The very first block from which a blockchain is constituted is known as the "genesis block" (**Figure 1**).

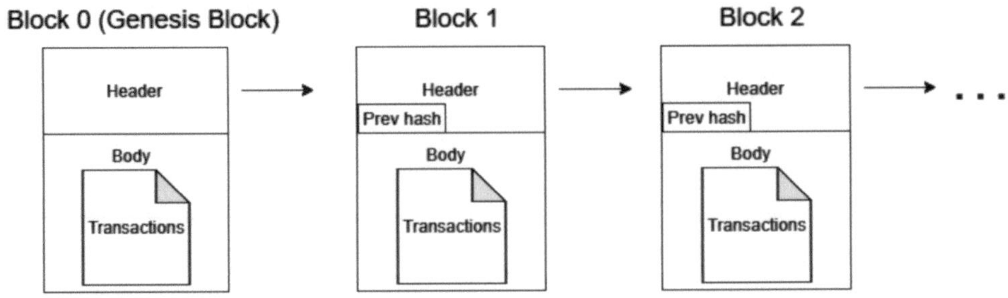

Figure 1.
Blockchain as a chain of blocks.

It should be noted again that a blockchain is a type of Distributed Ledger Technology (DLT) with a series of specific features. By DLT, we mean any type of technology that makes use of a distributed ledger and, therefore, not all DLTs are blockchains. As an example, new generation technologies, such as IOTA or Hashgraph, are based on DLT different from the blockchain, being named blockless technologies, which are out of the scope of this document.

As mentioned, in blockchain, the ledger is distributed between participants of a decentralized network without any central authority. In a public non-permissioned blockchain, all participants in the network keep a copy of the ledger, while in other more complex or restrictive kinds of blockchain, different ledgers can be held by subsets of participants. As an example of this statement, Hyperledger Fabric is presented as a permissioned blockchain technology, which allows us to separate the different nodes into different channels, having the nodes in the same channel the same copy of the ledger. At first sight, such kind of systems could be prone to issues related to the ledger synchronization. If any participant had the ability to promote their own version of the ledger and thereby their own version of the transactions, they could try to make a profit from it. However, how blockchain avoids this sort of incidences is through consensus mechanisms.

Consensus mechanisms govern the way participants storing and verifying blocks agree on one common version of the facts (a shared truth). The Consensus allows nodes to reliably validate new blocks in the network. There are a variety of proven types of consensus mechanisms, such as Proof of Work (PoW), Proof of Stake (PoS), Practical Byzantine Fault Tolerance (PBFT) or Proof of Elapsed Time (PoET), among other not-so-known ones, such as [2, 3], for example.

The most widely adopted consensus algorithm today is Proof of Work, used in both Bitcoin and Ethereum. Proof of Work basically consists of the resolution of a computationally complex challenge (related to the block itself) as a condition for the insertion of a block in the chain. The participants of the blockchain compete for the resolution of this challenge in return for a reward. The challenge is difficult to solve, but easy to verify so that the rest of the participants can easily verify the resolution of the challenge and agree on the new block. This algorithm guarantees consensus as long as no participant has more than half the computing capacity of the network, at the cost of high energy consumption. This high energy consumption and wastage of computing capacity is driving blockchain networks like Ethereum to migrate to lighter consensus algorithms, such as Proof of Stake.

The most used cryptographic function in Proof of Work is the hash. Hashes are trapdoor functions, which mean they are really easy to compute in one direction, but really hard in the opposite (find its inverse). When a participant of the network (called miner) finds a solution for a hash matching certain properties, it is enabled to assemble a new block and broadcast it. Upon reception, every other participant can efficiently check that the block is valid given that is linked to the last one and matches the properties required by the network. This validation can be computed efficiently due to hashes being trapdoor functions. The consensus is reached when every participant has the same blocks, in other words, every participant agrees on the chain composition (longest blockchain). Hashes are also key tools for verifying data integrity and for the cryptographic signature process.

All this said, what advantages do we get with the use of blockchain? What leads us to adopt a network with such a load of processing and redundancy? All this complexity is necessary to constitute a decentralized network composed of multiple participants that reach a common consensus without the intervention of a central authority; to build a transparent and immutable ledger verifiable by itself; to establish a contract without the intervention of a notary (in fact, applications running on a blockchain are known as smart contracts). And all these goals are achieved with the highest level of trustworthiness and availability. Of course, blockchain is not the solution to everything. It is not the right solution for systems governed by a single central authority or to store data whose integrity and source is not relevant. It is a new paradigm that ensures the deterministic execution of a contract and the incorruptibility of the data in a ledger with full guarantees and without the intervention of a third party.

More technical information has been presented by [4, 5] so that the reader can obtain further knowledge on the functioning of protocols.

2. Blockchain as a secure ledger

Once blockchain technology has been introduced, the focus is on the fulfillment of the information security properties it provides.

Focusing on data integrity, blockchain ledger is immutable. Every transaction in a block is cryptographically signed by its sender, every block in the blockchain is cryptographically signed by its miner, every block contains a hash of the immediately preceding block and all the participants in the blockchain network reach a consensus about the chain as the shared truth. To alter a single transaction in the blockchain, an attacker should alter each subsequent block accordingly, resolve the consensus challenge of that block and subsequent blocks, and persuade more than 50% of network participants to adopt the new chain. That situation is close-to-impossible, due to the hashing properties and the amount of computational and electrical power required to achieve this goal. Blockchain is tamper-resistant and integrity is the greatest of its merits.

Merkle trees are a fundamental use of hashing in blockchain technologies that have not been mentioned before in the article. Merkle tree summarizes all transactions in a block into a single fingerprint, allowing to verify that all transactions in the block have been included without modification. Below we can find an example of one of these Merkle trees (**Figure 2**).

As we can see above, each leaf in the Merkle tree is a hash of transactional data and hashing is applied recursively over each subset of hashes forming the tree structure. Merkle trees are not only applied to block transactions but sometimes also to the ledger state (the result of the execution of all ledger transactions).

Non-repudiation is another information security property intimately linked to integrity. Since every transaction in the blockchain is cryptographically signed by its sender and the chain is immutable, the sender can never deny having ordered the transaction. However, that sender, in general, cannot be associated with a physical entity, but only with an account (as we will explain when discussing about privacy).

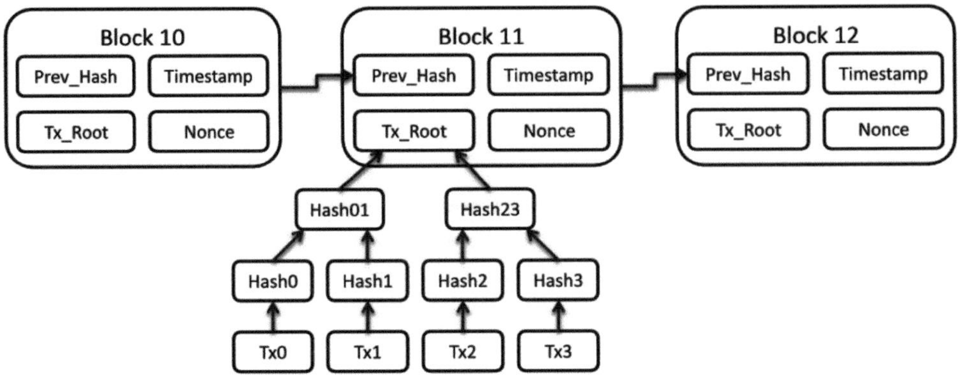

Figure 2.
Merkle tree.

In terms of availability, the distributed character of blockchain network makes it highly available. In addition, transactions on public blockchain networks usually involve a cost to the sender equivalent to their processing and storage consumption. This cost results in a reward for the miner of the block containing the transaction. Furthermore, it protects against Denial of Service (DoS) attacks, since an attack involves a cost proportional to the resources consumed for a potential attacker. For example, in Ethereum MainNet, this cost is reflected in the concept of gas. Gas represents the computational and storage cost of the transaction. At the same time, this gas has a variable cost in Ether, Ether that is obtained by mining or buying it. The availability concept is linked with the anti-SPoF (anti-Single Point of Failure) concept. Preventing a SPoF is usually a mandatory requirement when it comes to critical applications and, which need to offer a high availability rate, and even not-so-critical ones. If this point of failure is exploited, accidentally or intentionally by an attacker, the whole ecosystem breaks down, so it's interesting to be able to use resilient infrastructures, like Blockchain, to avoid this issue.

As for privacy, it is important not to confuse this concept with confidentiality, although they usually come hand in hand. In general, public blockchain networks bind transactions to accounts. These accounts are represented by a public-private key pair and may have a state associated with them, but they are not usually associated with an entity or individual. Only the individual in possession of the corresponding private key can launch a transaction on behalf of the account through a cryptographic signature, but the identity of the individual behind the key pair is unknown. In this way, a high degree of privacy is offered thanks to this pseudo-anonymity. Of course, there are identity management frameworks for blockchain, but these frameworks are not part of the core of a classic blockchain network.

One of the strong points of blockchain technologies is the transparency of transactions, a concept that in general is at odds with confidentiality (understood as encryption). Therefore, and except for specific blockchain technologies and private networks, blockchain does not provide encryption capabilities and this, if applies, must be implemented at the application level.

On the other hand, authorization is usually left to the application level in regular non-permissioned blockchain networks, while it can be part of the core of the technology in permissioned blockchain networks.

In short, we can conclude that blockchain is an extremely secure and resilient technology, but in general does not include confidentiality (understood as encryption) among its main objectives.

3. Blockchain for backup and recovery

Having shown to the reader the blockchain capabilities as a secure ledger, this section wants to analyze blockchain as a support tool to implement backup and recovery strategies. We have chosen this use case because it shows in a different way another use of blockchain, far from the common ones which usually appear in the literature.

One of the most innovative applications of blockchain technologies is to use it by secure storage and recovery systems. A Backup & recovery system usually has the following features:

- **Continuous/Automatic data backup**: It ensures that the changes you make to your files are simultaneously copied to the storage location. This lets you recover even the most recent changes in case of data loss, thus lowering your recovery point objective.

- **Incremental backup**: This is a type of backup where only the changes are copied, not the full file. This reduces the time taken for copying data and does not slow down your work.

- **Instant recovery**: This feature allows a backup snapshot to run temporarily on secondary storage to reduce the downtime of an application.

- **Data deduplication**: It eliminates duplicate data record blocks while data is transferred to the backup storage location. This reduces the network load and the storage space you require.

- **Error-free copy**: Data backup software features also ensure that the data copied from a source and stored at the backup server are the same and do not mismatch nor contain errors.

Historically, backup and recovery procedures were applied mainly to general-purpose devices in the enterprise environment. The number of incidents grows daily, and the consequences are increasingly alarming as, for example, security holes in IP cameras [6], DDOS attacks generated from the Mirai botnet [7, 8] known as Dyn Attack or event take control of a vehicle [9]. Due to these problems, Backup & Recovery systems are being extended to cover these devices too.

3.1 General-purpose devices

From the point of view of general-purpose systems, the main challenge that blockchain is expected to solve is the control data from tampering attacks; directly related to the integrity of the data.

We could find proprietary solutions that offer blockchain backup services at an enterprise level, see [10]. This solution provides mechanisms to ensure that legal documents existed on certain dates or to certificate authenticity of medical records.

3.2 IoT devices

Most IoT systems are managed through firmware so ensuring the integrity and authenticity of the firmware update of the devices is a complex and critical task that must be carefully addressed. In addition, it may happen that multiple devices with their various subsystems need to be updated urgently and simultaneously, for example, to apply a critical fix. Therefore, the high availability of updates is a requirement.

Most existing solutions for firmware upgrades depend on the client-server model in which the manufacturer delegates the firmware distribution process to the suppliers of its products. The central client-server architecture has the drawback to be a Single Point of Failure (SPoF), and in case the server is not available IoT devices cannot access resources (updates). There are two approaches: manual and automatic.

On the one hand, in the manual update process, the device owner must start the firmware update process. In general, this type of update is adopted by devices that have limited bandwidth or directly it is the owner who decides to do it this way. However, the manual firmware update mechanism is not as efficient as the owner of the device must perform all operations manually. In addition, there is a high probability that human error may occur during the firmware update process or that devices are outdated due to lack of resources for updating.

On the other hand, the automatic updating seems more tempting to be adopted today. This way, the manufacturer of the IoT device could initiate the firmware update without the active participation of the device owner. The current automatic firmware update process uses the client-server architecture, where the repository of the provider is the server and the IoT device becomes the client-side. In general, there are two ways to deliver the firmware from the server to the client: PUSH and PULL methods. The differences between these two methods are in the initiator of the project firmware upgrade process. In the PUSH method, the device manufacturer starts the firmware update process by distributing the firmware binary file. In the PULL method, on the other hand, it is the IoT device that starts the firmware update process by sending a binary request to download the firmware to the server.

In Ref. [11], a blockchain-based firmware check and firmware update was proposed for IoT device systems. In the Lee and Lee scheme, the blockchain technology is used in your firmware proposal to verify the firmware version and the firmware authenticity file, as well as to distribute the firmware binary to the nodes connected to the network. Each IoT device is a network node, so each node must store all or part of the chain in its local storage, which means that only a few IoT devices are able to adopt this solution. So, the Lee and Lee proposition is not suitable for a heterogeneous IoT ecosystem.

In Ref. [12] the application of blockchain technology was proposed to update the firmware of IoT devices from different vendors. In this solution, each IoT device must periodically probe any random node in the network to check the firmware version. When a device vendor publishes a new version of the firmware upgrade to the block network, the newly created firmware upgrade needs to be verified first by the network through a consensus protocol. When one of the IoT devices of the associated device vendor wants to perform the firmware upgrade process, the device must create a transaction for the firmware upgrade request. In this scheme, IoT devices would not be able to download the firmware from their corresponding vendor unless all nodes in the network have verified the associated firmware. In this solution, all network nodes must store all firmware that has been published on the network.

3.3 Distributed file system (DFS)

When we find use cases such as the previous ones that require a distributed storage it is necessary to resolve where to store the files and who can access them. Blockchain technology does not offer storage solutions and it is not a recommended practice to store files in the blockchain. A possible solution is the use of distributed storage systems, like the decentralized P2P file storage systems. When using this kind of storage, files are divided into pieces that are replicated in different peers. A peer requiring access to an archive collects pieces of this archive, which is partially located in several peers at a time. The performance is similar to that of the P2P BitTorrent network and files are indexed by their hash or fingerprint.

As the main solution for implementing this kind of storage is to use IPFS [13]. IPFS is a decentralized hypermedia P2P protocol that allows the storage of distributed files dividing the files in chunks and replicating them in the peers that require them. When a file is downloaded, chunks are collected from different sources at the same time. Each file is identified and accessed through its hash or fingerprint. IPFS is the basis of Filecoin, a distributed storage network based on Blockchain. This network basically integrates IPFS in a specific Blockchain network for data storage in which the nodes get tokens as payment for the storage service provided (and the customers pay them). As for privacy and access control, the IPFS protocol does not include any encryption mechanism or access control. It is up to the client or DApp to encrypt each file prior to sharing the archive to prevent its disclosure to third parties, which is not a very versatile and interoperable solution either.

In short, IPFS provides distributed and decentralized storage of large files with a certain degree of resilience, integrity, and very high availability. By storing in the Blockchain the hash of the files, which occupies only a few bytes, both systems are linked and the integrity of the file is guaranteed.

4. Blockchain and content delivery networks (CDN)

Another interesting use case, maybe not so known as the previous one, is the application of blockchain strategies to content delivery networks. These networks are widely used nowadays, so we have considered that they are a good example of how we can use blockchain to add value to existent processes or technologies.

4.1 Introducing the content delivery networks

A Content Delivery Network (CDN) consists of an overlapped network of computers containing different copies of the same set of data. The objective of its creation is to maximize the bandwidth available in a service to improve, as far as possible, the availability and access to data.

A client accesses one of the copies of the data. By providing information replicas and bringing closer the node that provides the service, the response time should be improved, and service outages avoided. But, how does that affect the information a customer can see? The Byzantine Generals Problem enunciated by [14] establishes that the components of a distributed computing system may fail, reaching a condition of imperfect information. In this situation, an observer could have different information depending on unnoticed facts, like the server consulted or the client's location. A different observer could have different information for the same service consulted if an inconsistent CDN state is making the network to fail in its responses. A consensus regarding which component has failed in the first place and which information is trustworthy would make things easier.

Prior to the emergence of Blockchain and the definition of the Distributed Ledger Technologies, it was already possible to find collaborative networks that allowed greater resistance to targeted attacks [15], such as DDoS. But it was difficult to incentivize a participant to offer their computing power to these networks. This lack of ability to attract new collaborators made the network growth very difficult and undermined the power of defense systems. Blockchain, as a new concept of distributed system, allows to give a reward to the participants who take part in the improvement of a security system.

In addition to its application in cybersecurity, it is also possible to find deployments of CDNs with other purposes such as databases and DNS services, either in private or in a collaborative way. But they can also offer other different services such as the exchange of multimedia files or the distribution of software.

As stated by [16], the distribution of services is thought of as a solution to the problem presented by a centralized service. The distributed nature of blockchain allows these services to be decentralized. The characteristics obtained are common to both approaches, of which the most important and their counterpart are listed below.

- The load on each individual server is lowered, but the number of servers of the system is increased.
- The network traffic is distributed, but the information needs to be synchronized.
- The latency is diminished, and the bandwidth increased, in exchange for a higher maintenance cost.

In short, the use of CDNs adds some advantages, but it also increases the complexity of the architecture. There are several aspects that are affected by the need for offering copy mirrors and closer access to the client. The original server must have substitutes to ensure the high availability of the service. On the other hand, it is necessary to ensure the consistency of the data served. As there are a number of geographically distributed machines, which theoretically have the same information at all times, synchronization problems may arise.

Additionally, there must be a constant internal routing service to find all nodes in the network, to synchronize information internally and to provide better customer service externally. Furthermore, all these mechanisms are based on a record of user accesses and server use that improves the quality of service but generates an additional cost in computing and storage.

4.2 Use cases

Usually, actors such as data centers, mobile operators, digital advertising companies or online music providers, act as clients for companies like ISPs, media or news agencies, which distribute their content using this type of system.

One well known and widely used example for distributed data management is the peer-to-peer exchange of *torrent* files. The BitTorrent protocol defined by [17] uses computer networks that simultaneously and in a decentralized manner upload and download content over the network. But these exchanges are made without order or agreement on what content is propagated. What if we established a mechanism for the verification and validation of the exchanges? What if in addition to data we could transfer value? What if each of these participants could execute a business logic accepted by all?

Cybersecurity is a fundamental aspect of the industry at a global level. In modern times much media attention is being given to attacks that appear and cause serious damage all over the world. It is curious that so many systems are affected by security breaches, because as [18] indicates the attack vectors have not changed in the last 20 years.

Although there are mechanisms for distributing content prior to Blockchain, all the defense systems offered by security companies are, to a greater or lesser extent, centralized. In contrast, attacks are distributed. This fact already places the defenders in an initial disadvantageous situation.

A Blockchain-based defense would behave like Uber or like carsharing: in these two examples, the goal is to take advantage of resources that are normally underused for most of the vehicle's useful life, whereas when it comes to blockchain, the goal is to be able to use the computation of a data center that is not being used at a specific time. Resources could be rented from other network members and used to manage a powerful coordinated defense system. All of this without affecting the other computer owners when they need to use their resources.

Notice that a Blockchain solution is intended to record changes of ownership, different states of information, etc. that happen between two or more parties. Both the origin and the destination are known, although in many implementations of Blockchain it is only pseudonymous. And the execution of each one of these changes is deterministic, meaning that it will end with the same result regardless of who executes it within the network.

Coming back to the BitTorrent example, the question is if it is possible to be sure that the content offered by another user will always be available and whether any user should offer me the same content. The answer is no. And this is what will be changed by using Blockchain.

4.3 The great leap

Blockchain has revolutionized the Fintech world as we know it today. Revolutionizing content distribution could be its next goal. The big bet is the decentralization of services and the suppression of a single trust entity, relying on the system operation on distributed services.

To leverage the Blockchain capabilities and create a CDN that is truly disruptive, a method has been sought to obtain good latencies, and also to allow p2p files to be exchanged securely, without requiring an external auditor.

Using Blockchain can improve fundamental aspects of computational efficiency. Businesses adopting Blockchain could save on infrastructure and gain greater flexibility in the services they offer. In addition, related aspects such as scalability, security, reliability and performance could be improved. But as explained above, Blockchain also requires a physical network, software, and security procedures to allow it to operate properly.

The method that will achieve the best result is the simplest in its conception. It consists of taking successful projects and arranging them in such a way that they work in an ideal flow. In other words, it is the creation of nodes that participate collaboratively in a large resilient network of file exchanges as in BitTorrent, using hash tables as explained in [19] about Kadmelia, and versioning the contents like Git. Everything self-certified by the network itself.

Storj, the before-mentioned IPFS, DECENT or BlockCDN are some of the initiatives that are based on the distribution of contents that Blockchain offers to create new horizons in the CDN ecosystem. These solutions take advantage of storage times, downloads or bandwidth to boost their businesses. This means that the creators of these systems, with very different market viewpoints, are able to encourage users to adopt the network that each one promotes. These networks are focused on the needs of the user and reward participants for maintaining the network, without the need for a trusted third party to intervene to control all of them.

This is how the concept of "distribution" is being reinvented in the Content Distribution Networks. Content transparency and user privacy begin a new path together.

5. Blockchain for threat intelligence

Another interesting use case for blockchain is threat intelligence. As written in [20], threat intelligence is an advanced process which involves gathering valuable insights including mechanisms, context, indicators, actionable advice and implications about an emerging or existing cyberthreat. Threat intelligence processes must be adapted to a company ecosystem to integrate it properly.

One of the issues related to threat intelligent these days is that companies usually spend a lot of time researching the same threats, while others are unnoticed. As a consequence, new tendencies emerge, being now crucial to be able to share information between different interested parties. Following this principle, different companies are able to share information about threats to benefit each other. In the end, a distributed ledger of shared information is the ultimate goal of the threat intelligence philosophy.

Decentralization in the threat management ecosystem is not new at all. Previous works, as [21], study decentralization strategies applied to threat intelligence use cases. Others, like [22], propose a shared infrastructure to implement a threat intelligence solution. With decentralization, a single view of data and information shared concepts, blockchain comes into mind. Synchronization between different parties is also a crucial requirement, which is naturally made by blockchain due to its peer-to-peer-oriented architecture, as stated before.

When discussing the application of blockchain for threat intelligence use cases, Smart Contracts are a good asset too. For clarification, a Smart Contract is a computer program shared between nodes in a network that can be executed by all of them with a deterministic output. This piece of code allows us to verify, enforce or perform specific actions that can be audited so everyone knows the logical flow of the system. In other words, everyone is aware of the system functioning and is enforced to comply with it. Furthermore, the consensus is presented as a mechanism to guarantee synchronization between all the nodes. The aforementioned Smart Contracts enable high-level computations far from traditional distributed architectures focused on only-sharing information. In addition, we can ever think more philosophically and say blockchain is a more futuristic solution due to the fact that it allows us to create networks controlled by no-one, but verifiable by everyone.

As an example, specifically focusing on healthy ecosystems, a European initiative is trying to implement a blockchain-based Threat Management platform, which is the SPHINX Project [23]. In this project, health IoT devices within different medical centers share information about different threats ideally affecting the same ecosystem. Different components, within the scope of the same project, read from the same registry, so all of them have a single view of the data. This is one step forward in decentralization and information sharing solving a very specific problem applied to a very specific scenario. Focusing on the blockchain infrastructure, it acts as a BaaS (Blockchain as a Service), whose nodes are in different medical centers and the different IoT devices act as the users of this shared platform. This is a very clear example of how we can use Blockchain to solve a threat management problem in a wise way.

On the other hand, when it comes to other general cybersecurity solutions, blockchain can add some additional value to the traditional systems. For example, a very interesting use case is the distributed intrusion detection systems. However, these distributed intrusion detection systems are far from being fully secure as shown in [24], where the authors study the vulnerabilities that affect these systems. Blockchain can work as a distributed intrusion detection system, as shown in [25], avoiding the need to trust in third parties. It can also be very useful to detect some zero-days attacks in industrial environments by doing what we have named "log comparison", which basically consists of comparing different logs from different devices against the ones stored in a Blockchain infrastructure. When an attacker breaks into a system, one of the first things he usually does is to delete every proof of his presence, so he usually tries to delete every log which can link him with a particular incident. By having a trusted anti-tampering infrastructure, we can detect almost in real-time if a system has been compromised or not just comparing the logs in the system with the ones stored in the Blockchain, which are immutable "by design". It is important to mention that Blockchain grows very fast in disk, but storing just simple information, like log hashes, for example, we can easily overcome this issue.

No just focusing on pure threat intelligent, rather than monitoring activities, there are some studies which apply blockchain to enhance logging systems. One of the first examples is [26], written by some members of the University of La Sapienza in Rome and the University of Southampton, tries to find a solution to the European project Sunfish based on a distributed database which provides integrity and stability to the data, analyses the advantages and disadvantages of using this tool by implementing cloud computing. Nokia Bell Labs published a small report [27] in which it proposes to make use of private and permissioned blockchains instead of public ones to manage the logs, in this case, it focused on information related to banks. As mentioned in the paragraph before, storing logs can be problematic. As a consequence, working with hashes is wiser, because it is always possible to get the integrity of the data without affecting blockchain the disk usage excessively.

To sum up, blockchain comes up when sharing information between different parties is a matter. Whether if we want to identify the issuers of this information or if we want to anonymize them, different blockchain technologies can help us to achieve these requirements.

6. Conclusions

As we have read, blockchain is much more than just cryptocurrencies. It is possible to build a vast number of use cases by using blockchain as a trusted infrastructure due to its security properties. In this document, we have shown several of these use cases, all of them security-related may be unknown for the reader and different from the now-trendy cryptocurrencies trading.

As far as we dig into the blockchain technology, we become more aware of its possibilities, ranging a huge spectrum of functionalities and covering various use cases in different fields, such as industry, health, finances... although this document has enlightened only the ones concerning the security field.

However, the future is continuously changing, and blockchain technologies are not the panacea for every problem in the world. The emergence of the so-called blockless technologies is a challenge for the blockchain technology itself, because they present a different way to achieve almost the same security requirements of the blockchain technologies, but trying to overcome its issues, such as latency and fees. The subsequent years will decide which ones of these technologies take advantage of the rest of them, but the decision does not seem to be easy.

Acknowledgements

This work was performed with the financial support of the ELKARTEK 2018 (CyberPrest project, KK-2018/00076) research program from the Basque

Government. At the same time, this content is the product of a joint effort of a group of people belonging to the Tecnalia Blockchain and Cybersecurity Research Group and it is the result of our experiences in researching, developing and applying blockchain to different sectors. We want to thank the community of hard-working developers involved in foundations and technologies like Hyperledger or Ethereum, among others, allowing us to collaboratively improve and develop new blockchain-based solutions to reach a better world.

Author details

Oscar Lage*, Santiago de Diego, Borja Urkizu, Eneko Gómez and Iván Gutiérrez
TECNALIA, Parque Científico y Tecnológico de Bizkaia, Derio, Spain

*Address all correspondence to: oscar.lage@tecnalia.com

References

[1] Nakamoto S. Bitcoin: A Peer-to-Peer Electronic Cash System. 2008. Available from: http://bitcoin.org/bitcoin.pdf

[2] Innerbichler J, Damjanovic-Behrendt V. Federated Byzantine Agreement to Ensure Trustworthiness of Digital Manufacturing Platforms. 2018. pp. 111-116. DOI: 10.1145/3211933.3211953

[3] Fan X, Chai Q. Roll-DPoS: A Randomized Delegated Proof of Stake Scheme for Scalable Blockchain-Based Internet of Things Systems. 2018. pp. 482-484. DOI: 10.1145/3286978.3287023

[4] NRI. Survey on blockchain technologies and related services [Tech. Rep.]. 2015. Available from: http://www.meti.go.jp/english/press/2016/pdf/0531_01f.pdf

[5] Zheng Z, Xie S, Dai H, Chen X, Wang H. An Overview of Blockchain Technology: Architecture, Consensus, and Future Trends. 2017. DOI: 10.1109/BigDataCongress.2017.85

[6] "Webcam Maker Takes FTC's Heat for Internet-of-Things Security Failure" [Internet]. 2013. Available from: https://www.technewsworld.com/story/78891.html

[7] Hilton S. "Dyn Analysis Summary Of Friday October 21 Attack." 2016. In: Oracle Dyn Company News. Oct 26. Available from: https://dyn.com/blog/dyn-analysis-summary-of-friday-october-21-attack/

[8] Chacos B. Major DDoS attack on Dyn DNS knocks Spotify, Twitter, Github, PayPal, and more offline. 2016. In: PCWorld [Accessed: 22 October 2016]

[9] Bonderud D. Eight Crazy Hacks: The Worst and Weirdest Data Breaches of 2015 [Internet]. 2015. IBM SecurityIntelligence December 9, 2015. Available from: https://securityintelligence.com/eight-crazy-hacks-the-worst-and-weirdest-data-breaches-of-2015/ [Accessed: 05 March 2019]

[10] "Stonefly: First Blockchain Technology Backup" [Commercial tool]. Available from: https://stonefly.com/blockchain-backup

[11] Lee B, Lee JH. Blockchain-based secure firmware update for embedded devices in an internet of things environment. The Journal of Supercomputing. 2017;**73**(3):1152-1167

[12] Boudguiga A, Bouzerna N, Granboulan L, Olivereau A, Quesnel F, Roger A, et al. Towards better availability and accountability for iot updates by means of a blockchain. In: 2017 IEEE European Symposium on Security and Privacy Workshops (EuroS&PW) (pp. 50-58). IEEE; 2017

[13] Benet J. "Ipfs-content addressed, versioned, p2p file system". 2014. arXiv preprint arXiv:1407.3561

[14] Lamport L, Shostak R, Pease M. The byzantine generals problem. ACM Transactions on Programming Languages and Systems. 1982;**4**(3): 382-401. DOI: 10.1145/357172.357176. Archived from the original on 13 June 2018

[15] Castro M, Liskov B. Practical byzantine fault tolerance and proactive recovery. ACM Transactions on Computer Systems. 2002;**20**(4):398-461

[16] Domenico T, Trunfio P. Toward a synergy between p2p and grids. IEEE Internet Computing. 2003;**7**(4):96-95

[17] Cohenv B. Incentives build robustness in BitTorrent. In: Workshop

on Economics of Peer-to-Peer Systems. Vol. 6. 2003

[18] Shinde PS, Ardhapurkar SB. Cyber security analysis using vulnerability assessment and penetration testing. In: 2016 World Conference on Futuristic Trends in Research and Innovation for Social Welfare (Startup Conclave). IEEE; 2016

[19] Maymounkov P, Mazieres D. Kademlia: A peer-to-peer information system based on the xor metric. In: International Workshop on Peer-to-Peer Systems. Berlin, Heidelberg: Springer; 2002

[20] Shahare R. "Blockchain, for Threat Intelligence Maybe?" [Internet]. 2019. Available from: https://www.cpomagazine.com/cyber-security/blockchain-for-threat-intelligence-maybe/

[21] Burger EW, Goodman MD, Kampanakis P, Zhu KA. Taxonomy Model for Cyber Threat Intelligence Information Exchange Technologies. 2014. pp. 51-60. DOI: 10.1145/2663876.2663883

[22] Wagner C, Dulaunoy A, Wagener G, Iklody A. MISP -the Design and Implementation of a Collaborative Threat Intelligence Sharing Platform. 2016. DOI: 10.1145/2994539.2994542

[23] "SPHINX - A Universal Cyber Security Toolkit for Health-Care Industry". 2018. Available from: https://cordis.europa.eu/project/rcn/220226/factsheet/en [Accessed: January 2019]

[24] Li W, Meng Y, Kwok LF, Ip HHS. PMFA: Toward passive message fingerprint attacks on challenge-based collaborative intrusion detection networks. 2016;**9955**:433-449. DOI: 10.1007/978-3-319-46298-1_28

[25] Meng W, Tischhauser EW, Wang Q, Wang Y, Han J. When intrusion detection meets blockchain technology: A review. In: IEEE Access. Vol. 6. 2018. pp. 10179-10188. DOI: 10.1109/ACCESS.2018.2799854

[26] Gaetani E, Aniello L, Baldoni R, Lombardi F, Margheri A, Sassone V. "Blockchain-based database to ensure data integrity in cloud computing environments". 2017. Available from: eprints.soton.ac.uk

[27] Shekhtman LM, Waisbard E. Securing log files through blockchain technology. In: Proceedings of the 11th ACM International Systems and Storage Conference (SYSTOR '18). New York, NY, USA: ACM; 2018. pp. 131-131

Chapter 7

Blockchain: From Industry 4.0 to the Machine Economy

Oscar Lage

Abstract

The extreme automation of our factories is necessary in order to face the Fourth Industrial Revolution. This new industrial paradigm will force our industries to manufacture much shorter and customized series at increasingly competitive prices, even tackling the manufacture of thousands of different configurations of a single base product. In order to achieve this, our production processes must have a flexibility in their configuration that has never been imagined before. This flexibility and ability to adapt automatically to demand are the essence of the Fourth Industrial Revolution and are part of the Western strategy to recover an industrial sector increasingly threatened by the Eastern production of large series at really competitive prices. Based on our participation in more than a dozen proofs of concept in the automotive, aeronautics, agri-food, or energy sectors, we describe the scenarios in which blockchain technology brings the greatest benefits to Industry 4.0. After finishing different experimentations, we carried out an in-depth analysis of the true added value of blockchain in the industry and contrasted our conclusions through interviews with more than 20 people in charge of innovation from different industries. As a result, we have obtained the principal four values of blockchain technology applied to Industry 4.0.

Keywords: blockchain, DLT, Industry 4.0, trust, cyber security, IoT, IIoT, industrial systems

1. Introduction

The automation of our industries and the relationships of the different agents in the value chain will allow us to eliminate many repetitive manual processes with little added value that reduce the competitiveness of the industry [1]. Even the automation of tendering and contracting processes can improve our competitiveness.

Technologies, such as artificial intelligence, flexible robotics, IoT, or augmented reality will allow us to advance in the digitalization and optimization of our processes, but the great barrier to implement a fully automated production systems and especially relationships is precisely the lack of trust and security [2].

Trust is the basis of a new research line that in recent months has had an increasing impact on industrial forums and conferences: blockchain technology. Blockchain is a distributed ledger of transactions and digital events that have been executed and shared among participating parties. Each transaction is consensuated, mathematically linked and stored by the network of participants, thereby achieving its immutability. Blockchain allows us to operate our procedures and relationships in the digital environment in a much more safe and reliable way [3].

The next few years will see a profound transformation of industrial processes, increasing the synchronization between different agents in the value chain, as well as extreme automation of decision-making, all thanks to the reliability offered by blockchain. It is even hoped that in the future, it will be able to transform its own business models, just as in recent decades the Internet has done, which has so far been the most disruptive technology in history.

In this chapter, we will explain the different use cases and scenarios that we consider to have greater potential in the future of Industry 4.0, starting first with generic industrial cases and then analyzing the specific cases of the energy industry. This selection has been made based on the experience of more than a dozen blockchain projects in the domain of Industry 4.0.

Next, we will describe the four main generic values that we have discovered after different proofs of concept with several companies. Finally, we will discuss future lines of research linked to a new concept such as the machine economy and report the final conclusions of the chapter.

2. Bringing blockchain into Industry 4.0

After carrying out different proofs of concept, mainly associated with manufacturing companies, as well as analyzing other experiments carried out by third parties, we expose in a critical way which would be the main application scenarios of the blockchain technology and its benefit for industrial companies.

All the analyzed cases have been contrasted through a working meeting with several companies in order to analyze the real need and utility of them. The following are the use cases that have presented greatest utility in the experimentation process, responding to real needs not covered today in their ecosystems.

2.1 Traceability

The traceability of industrial goods throughout the entire supply chain, including even the life cycle of a product, is one of the applications that according to consulted experts in the experimentation, as well as the level of maturity of the technology in this field, is expected to have a greater impact on the short/medium term of the industry.

For any point in the chain, it is very valuable being able to have visibility of the destination and use of its components; thanks to this information the participant in the supply/value chain will be able to (i) analyze the impact of any change in the design/composition of their product, (ii) anticipate changes in consumption habits/trends, (iii) avoid manually entering details of the products/components received by suppliers, (iv) automate complaints and warranties without the need for paperwork, or even (v) avoid reusing certificates of origin.

We are facing a known need that the big industrial players have wanted to solve on different occasions [4–7]. The large industries have designed and built traceability systems based on traditional (centralized) architectures and have made them available throughout their sectorial supply chain. However, these systems have not been widely accepted, and the only ones that continue to exist are those related to food safety that is mandatory.

The problem with the previous approaches is that the "giant" of the supply chain was the one that offered its system to the rest and was in charge of the custody and coherence of the common database.

This created great reticence because, even if industrial data visibility policies were implemented so that only agreed users/companies could consume certain information, there was a "demigod" in the supply chain which, due to the architecture of the system, could have visibility and exploit the information of the entire value chain. Furthermore, processing the information in a traditional system is very complicated to guarantee the sovereignty and protection of industrial data [8].

The alternative to create a similar system using traditional technologies is to create a clearing house in the supply chain, which has been done in areas such as food safety and is the only area where traceability is complete throughout the chain [9]. However, in this case the actors only submit information related to food safety and cannot consult/exploit the information, so the functionality is not full.

Blockchain makes it possible to eliminate these barriers thanks to a distributed architecture in which there is no "agglutinator" of the contents. Guaranteeing through "contract" and cryptography the visibility and use of data (sovereignty of industrial data) and ensuring that all participants in the network are treated equally.

However, we have detected that an important point in these projects is to maximize and automate as much as possible the capture of data, which is why industrial projects are considering that the Industrial Internet of Things (IIoT) should be the origin of most of the data that are dumped in the traceability chain. Moreover, this information should be signed by means of cryptographic hardware in these IIoT devices, so that the reliability of the data would be extraordinary.

2.2 Interoperability and sovereignty of industrial data

Data and its exploitation are going to be the key in this new industrial paradigm in which we are entering, promoting even service models based on data [10]. That is why it is said that data is the new industrial raw material and its sovereignty is a key point today.

For this reason, several initiatives have arisen that could be called industrial data platform and that aim to manage and share data of industrial processes, as well as create value-added services based on them. The most evolved platforms, such as the one from the international data space consortium, which arose in Germany but is currently the leading European experimental platform, even include application/service marketplaces based on industrial data [11].

Perhaps predictive maintenance together with other cases of data analysis and prescription are the most common and tangible cases today [12], but it is expected that really these platforms are the basis for innovative proposals of business models and industrial services that today we cannot even imagine. However, there is currently a major barrier to the adoption of such platforms, and again it is the reliability of the industrial data and its protection.

Firstly, there are models for selling information related to industrial processes, the value of which will depend on the reliability of such data. Therefore, it is one of the reasons why blockchain begins to be a buzzword in the deliberations on the future of these platforms, since the more reliable the data, the greater will be its value in the market.

On the other hand, these platforms must guarantee the sovereignty of industrial data, for which blockchain architectures/platforms that natively allow confidentiality between parties seem the most promising [13]. Current developments include data encryption models specific to each recipient or set of recipients, such as channels or private data collection in Hyperledger Fabric v1.4.

However, blockchain and smart contracts will even allow to execute algorithms and data processing independently, offering the recipient only the result of its execution [14]. In the future the algorithms can be encoded in a native blockchain program—the smart contract—in such a way that the owner of the algorithms can allow the smart contract to access and process their data and generate insights about them. However, the smart contract provider will not have access to the user's RAW data; this will allow them to offer a service based on the data without the customer having to make a disclosure of such information [15].

After all, it will allow us to put in value the industrial data even without having to expose them to a third party, allowing them nevertheless to execute certain processes on them. This can even be very useful to test/train prediction models of all kinds without endangering the source data, the result of which can then be a high-value algorithm for a specific industry.

2.3 IIoT reliability

One of the main benefits of the blockchain application to IIoT in which all the interviewed experts agree is precisely the decentralized architecture that blockchain can offer to IoT in general and especially to the industrial ecosystem whose requirements are more severe [16].

Currently the architecture of these systems is a classic client/server, which has a series of barriers and deficiencies for an environment such as IoT/IIoT. It is expected that the client/server architecture will not be able to respond to the exponential growth of IIoT and IoT in general; we must bear in mind that we will face an immense number of devices generating and consuming information from third parties. To get an idea of this figure, an industrial control machine or device generates hundreds of millions of data/parameters annually, and inside a medium-sized factory, we can find tens or hundreds of devices.

The cost of centralized processing and even network equipment and connectivity to support such cross traffic between different industrial systems (clients) with dependencies between them would be exponential if all these communications had to pass through a central system (server). In addition, this central system (server) would be a major bottleneck for all connected devices and a single point of failure (SPOF) which, if compromised, could generate a production shutdown of millions of euros in a single factory.

The trend is also that connected machines and factories interact outside their business environment with partners, suppliers, and customers. This brings another set of challenges at the level of identity management and device authentication. Currently within a factory, existing systems have multiple limitations because vendors deploy centralized systems that cannot interact safely and reliably with third parties, even rely on costly and complex in-house or manufacturer-controlled PKI architectures. In a global economy and in an ecosystem relationship, the problem and complexity multiply. Thus, blockchain technology has demonstrated that distributed authentication and identity management are highly efficient and feasible [17] and can solve identity management problems.

For all these reasons, we are dealing with a new paradigm in which, after moving from the traditional server model to an elastic cloud server architecture, we must evolve toward a network of devices in which blockchain is postulated as the main technological enabler. This paradigm shift would lead us toward decentralized registers that could become sectorial or even universal.

But the adoption of blockchain in the IIoT ecosystem, and IoT in general, offers another series of advantages, which although perhaps less disruptive also resolves some of the challenges and barriers to adoption of IIoT and IoT discussed above.

Blockchain offers us a decentralized record of information, which is also reliable and unalterable. That is why besides avoiding the single point of failure of traditional systems, it offers us a more resilient system, not only in terms of system availability, which increases exponentially by avoiding the single point of failure, but also in terms of information, since it provides us with a reliable record [18].

Offering a reliable record of information due to its immutability and ensuring non-repudiation of operations are an enabling factor for transactions between unknown devices or different organizations.

As we have mentioned before, one of the biggest barriers to adopting a higher level of automation in the industrial environment is precisely the mistrust of data, especially data from third parties. Although the industries themselves in many cases do not rely on automating some critical processes based on their own information due to potential sabotages or failures, it is impossible to think that they will do it based on third party information sources.

Blockchain offers reliability over our own information—thanks to the integrity and strong authentication of our issuers—as well as over information provided by third parties. Such reliability will allow greater automation and avoid many of today's low value-added manual processes that are provoked by a lack of confidence in the data.

The decentralization of information and its immutability are also a major advantage for critical industrial infrastructures (chemical, energy, etc.). According to the latest recommendations for critical infrastructure protection like the European Critical Infrastructure Protection (ECIP) or NIST Cybersecurity Framework, they should be able to guarantee the custody of their data in the case of any fortuitous incident (natural disaster, system failure) or deliberate incident (physical and/or logical attack) for forensic analysis.

Nowadays, this custody of information in case of cyber incidents is practically impossible to achieve since the attacker usually stays inside the system 146 days before executing the attack or being detected [19], and one of its objectives is to meticulously study the infrastructure not only to maximize its impact but also to be able to erase any trace once the cyberattack is executed.

This is why traditional backup systems and data replicas are usually eliminated during the attack; however, if the infrastructure was connected to a blockchain network, the attacker would have to completely erase each and every one of the nodes of the distributed blockchain network to make their footprints disappear, something totally unthinkable. In fact, during all the time that the attacker remains investigating, the infrastructure is erasing his trail, so a simple periodic comparison of the logs of the infrastructure itself against its unalterable copy in blockchain could alert us of the existence of an intruder in the network or detect any change in the machine code of our industrial devices.

However, although blockchain is postulated as the solution to IIoT's architectural design problems, it must be kept in mind that current solutions and ledgers must evolve in order to respond to the needs of IIoT devices in real time (low latency, bandwidth, message size). That is why in the blockchain, ecosystem begins to emerge new developments and technologies aimed at overcoming this barrier [20–22]. If this is achieved, the potential market and technological impact could lead to the long-awaited paradigm shift we were talking about earlier.

3. A new energy industry

In the last years, the energy sector has initiated a major transformation of the electricity grid, the industrial infrastructure responsible for transporting and distribution electricity from the generation plants to the consumer. The smart grid is a much more automated and resilient grid and offers unprecedented levels of reliability and service continuity.

3.1 Energy sector considerations regarding the previous section

The smart grid itself is a network of IIoT devices and is also considered a critical infrastructure, so everything mentioned above about the advantages of using blockchain in IIoT devices obviously applies directly to this industry.

Traceability is also relevant in the energy industry; therefore, at the end of 2018 ACCIONA announced, in collaboration with Tecnalia, the first proof of concept for the use of blockchain to trace the renewable origin of energy. In this case the fundamental objective of traceability is to guarantee the renewable origin of the energy and thus differentiate the energy generated in a sustainable way.

Even so, since the initial experimentation, there are several utilities that have made different proofs of concept, and we must distinguish between (i) the traceability of energy from its point of origin, with information collected from the IIoT itself (smart meters of the power plant) or (ii) the traceability made retrospectively based on the data that the utility itself (not the machines) introduces in the blockchain. The first one gives a total guarantee and trustworthiness; in the second case, the reliability is given by the utility itself and does not have a superior value than a report signed by the energy company itself.

Equally important is the interoperability and sovereignty of the data in a smart grid in which different operators and manufacturers collaborate with a common industrial objective—the grid resilience—but with competing business objectives.

3.2 Prosumers and the value of energy data

We are facing a decentralization of energy production in part due to a new participant in the ecosystem, the prosumer [23]. Prosumers, unlike a traditional consumer—who simply consumes the energy provided by the smart grid—also are able to produce its own energy (**Figure 1**).

The proliferation of prosumers in the energy ecosystem is going to cause that these consumers will have more information and detail than the utility itself, something unthinkable until now where every kilowatt consumed by a home or company is accounted by the energy distributor.

These prosumers may be consuming energy without the utility being aware of it, but they must provide service to the user if it punctually needs more energy than is able to produce, either because of an increase in consumption, because the user has photovoltaic generation on the roof but the day is cloudy, etc.

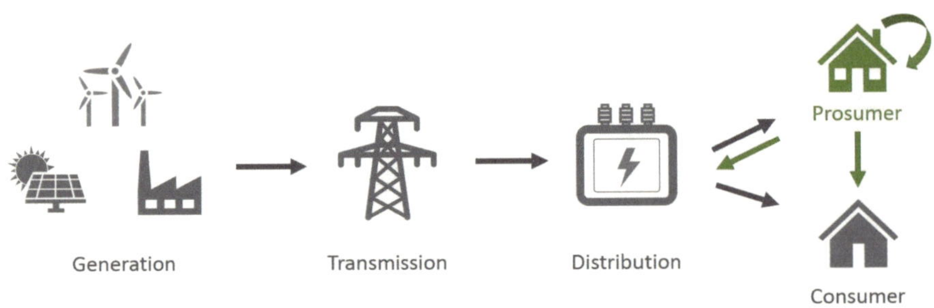

Figure 1.
Smart grid architecture and energy flows including prosumers.

In fact, these users have critical information to operate the system that will be extremely valuable for the stakeholders of the energy system in order to optimize their processes and ensure the stability of the network. It will allow them also to predict energy demand more accurately, avoiding deviations in the daily markets, improving the balance of the grid, and so on. Even in the case of large consumers, some companies offer optimized energy savings based on a baseline measurement.

However, the user is increasingly aware of the value of these data and not only because of their impact on the energy ecosystem. Starting from the detail of energy consumption, it is possible to infer a quite exhaustive profile of the user and, for example, to carry out a very good segmentation for marketing impacts.

The following transformation of the energy sector could be precisely based on the exploitation of these data, and thanks to blockchain, users could have control of them and therefore of their privacy.

4. The core value of blockchain in the industry

After analyzing the results of different proofs of concept and the benefits provided, we could say that blockchain can bring a number of differential features to Industry 4.0.

Perhaps the most popular is the decentralization of processes and business models. Blockchain provides by definition the intermediation between two parties in a reliable way [24] that is why many processes and organizations whose main value is the intermediation between parties can be optimized thanks to blockchain technology. We will therefore see intermediaries that adopt technology to be more efficient and robust, thus being able to offer a better service at more competitive prices or consortiums of companies that invest in creating themselves platforms to manage their relationships without depending on current intermediaries.

At the same time, blockchain offers an unalterable record of the history of any asset or industrial good, so traceability on that record is natural for blockchain technology. In addition, this record can be shared with third parties in an exercise of transparency of their processes.

Blockchain offers a really efficient synchronization of processes; it provides us with a single consensuated vision of the information related to industrial assets and processes, something really important in cases where different players and information systems must be coordinated to achieve a common industrial objective.

It is a perfect synchronization technology, resilient to network microcuts or failures of the systems involved in the industrial process. These usual deficiencies of the traditional technologies generate incoherencies in the data and consequently incorrect decision-making due to a bad synchronization of the information shared between the collaborating systems.

Finally, we should emphasize the blockchain capacity for process automation thanks to being a reliable source of information by offering a synchronized, consensual, and unalterable record on which we can also have a non-repudiation of the information, as each participant signs each of their transactions as if it were a digital contract in pdf.

As we have already commented, automating our industrial processes based on information from third parties is really risky if the source is not reliable. Unlike the technologies that we usually handle, blockchain offers us that certainty, even an evidence that can be used to claim a third party if the recorded information is not real or accurate.

5. Machine economy

The previous sections focused on explaining the results of proofs of concept and analysis of the applicability of blockchain in Industry 4.0, mainly in the improvement of processes and the creation of new products/services. In the current section, the focus will be to introduce a new economic paradigm that arises from the merger of industry, economy, and disruptive potential of blockchain, an area that precisely because it is still very experimental opens different lines of future research, the machine economy.

To understand the machine economy, we must first understand how we are facing a new paradigm of decentralization and disintermediation, which is already a small phenomenon in the world of currencies and will soon be a reality in many other areas. Entities such as eBay or Amazon already have to face the competition of OpenBazaar, an open-source blockchain software that offers near the same value as those companies. At the same time, the highly appreciated platform business models such as AirBNB or UBER are reflecting on what value to contribute beyond intermediation; otherwise they will be disintermediated by blockchain technology.

But the real potential of blockchain is not just to eliminate intermediaries; really these "cryptocurrencies" are digital tokens that represent a value [25]. Obviously the simplest application has been to create cryptocurrencies in which the blockchain issued those tokens instead of a central bank, but those tokens can represent whatever we want. Those tokens can represent the possession of a house or the identity of a person and all their history, but they can also represent the right to consume a service, to make decisions about the future of an organization, etc.

And this is where the real disruptive change begins; with the so-called crypto economy or token economy, an economy dominated by these tokens that is cryptographically protected by the blockchain will change the rules of the game and allow the total decentralization of the economy. In this new economy, the value will be tokenized, and these tokens will represent very different values as we commented.

This token economy is already emerging, it started with the cryptocurrencies, and we have also lived a new paradigm in the search for funding for business projects, in which under the name of initial coin offering (ICO) entrepreneurs with disruptive ideas find a new blue ocean of funding [26–28]. These entrepreneurs sell tokens that in many cases represent a service of that startup in the future, something similar to crowdfunding but totally globalized and without intermediaries who must manage those rights of future use of a platform. But these projects are going one step further than a simple decentralized crowdfunding; they are even devising new types of autonomous and decentralized organizations known as decentralized autonomous organizations (DAO) [29].

These organizations are created and financed by the community in order to offer an autonomous service thanks to blockchain. Imagine that we are tired of Google, Twitter or Facebook continues to earn money with our personal data, but we do not want to lose its functionality. Blockchain allows the community to finance and launch a new social media, or any other service, but without being managed by any for-profit entity, nor has a company registration number (CRN) in any country. It will be a virtual organization offering the service and relying on the community to perform those tasks that cannot perform by itself as investment decisions or strategy. So the community itself will run this virtual organization in a format similar to how a federation of worker cooperatives works.

This organization will be able to charge for its services and reinvest all the benefits in the development of improvements, new functionalities, etc. These organizations could also share part of those benefits with their promoters and community or simply offer these users free services. In this type of organizations, the "shareholder pact" has been programmed since its creation, "code is law." In fact, the change of these rules will have to be agreed by the community of users.

Machine economy is precisely to transfer this concept of DAO to the machines; we could be in front of a new evolution of the IoT. Let us imagine, for example, something we all know, a car. In a few years, it would not be difficult to imagine that there are a significant number of users who do not have a car and that there is a fleet of cars at their disposal.

These cars could be sovereigns; they could have their own identity, history, and even their own "wallet" to store digital value (tokens) that they will use to manage and store the value they receive by offering their transport services to passengers, as well as to pay for their recharges, tolls, cleaning, and maintenance.

In this way, we turn this car into an economic agent itself, with its own economy, self-sufficient, and even with its own business model. What's more, this car would foster new micro-service ecosystems around it.

Let us go a little deeper into tokenomics and the machine economy. These cars could be offered by a company, in a similar way to the traditional model. But thanks to blockchain, this could be financed as a kind of crowdfunding in which a DAO would be created with the initial investment, and gradually it would increase the fleet, grow geographically, and even replace old vehicles. The DAO would also be able to offer truly affordable costs to its customers and allow token owners governance, decision-making, and profit-sharing.

In this way, transport could be outsourced to the machines; the same outsourcing exercise could be carried out to other machines—robots—for the washing of these cars, their maintenance, carried out by robots and even the printing of parts on demand, the rubbish collection service, etc.

The token economy aims to return the power to the citizenry, and thanks to being a fully digital economy, machines can be active agents of it, thus generating their own economy, the economy of machines.

However, nowadays the machine economy is mainly an experimental concept that requires solving different challenges. Some of these research challenges are (i) secure hardware-based digital identity, (ii) interoperability an data sovereignty, (iii) more scalable and computationally efficient DLT architectures, or (iv) distributed machine governance model, between others.

6. Conclusions

In this chapter we have analyzed the general applicability of blockchain technology to the new paradigm of the Fourth Industrial Revolution, and due to its particular peculiarities, we have made a brief analysis of the specific case of the energy sector.

Based on our analysis and experimentation, we have selected three main lines of generic application for Industry 4.0: (i) traceability, (ii) interoperability and sovereignty of industrial data, and (iii) IIoT reliability. Moreover, in the case of energy, beyond exposing any particularity linked to IIoT or energy traceability, the analysis has focused on the prosumers and the value of their data in a new decentralized energy ecosystem.

As an outstanding contribution, the conclusions on the real value of blockchain in the industry should be pointed out, where abstracting from any specific scenario, the value of blockchain technology in this sector is analyzed in a universal way. The results are four main values of the technology, which in addition to being really the core of the analyzed cases could become applicable in other sectors. These differential features can be very useful to detect in an agile way if the application of the blockchain technology in a project contributes with a differential value in front of the rest of technologies of the state of the art.

Finally, we end with a reflection on a new paradigm that we have discovered during our research, and that may open different lines of future research, the Machine Economy.

Acknowledgements

This work was performed with the financial support of the ELKARTEK 2018 (CyberPrest project, KK-2018/00076) research program from the Basque Government.

Author details

Oscar Lage
TECNALIA, Parque Científico y Tecnológico de Bizkaia, Spain

*Address all correspondence to: oscar.lage@tecnalia.com

References

[1] Schuh G, Potente T, Wesch-Potente C, Weber AR, Prote JP. Collaboration mechanisms to increase productivity in the context of Industries 4.0. Procedia CIRP. 2014;**19**:51-56

[2] Marques M, Agostinho C, Zacharewicz G, Jardim-Gonçalves R. Decentralized decision support for intelligent manufacturing in Industry 4.0. Journal of Ambient Intelligence and Smart Environments. 2017;**9**(3):299-313

[3] Drescher D. Blockchain Basics. Berkeley, CA: Apress; 2017

[4] Banterle A, Stranieri S. The consequences of voluntary traceability system for supply chain relationships. An application of transaction cost economics. Food Policy. 2008;**33**(6):560-569

[5] Maro S, Steghöfer JP, Staron M. Software traceability in the automotive domain: Challenges and solutions. Journal of Systems and Software. 2018;**141**:85-110

[6] Farris M, Wittmann C, Hasty R. Aftermarket support and the supply chain: Exemplars and implications from the aerospace Industry. International Journal of Physical Distribution and Logistics Management. 2005;**35**(1):6-19

[7] Heyder M, Theuvsen L, Hollmann-Hespos T. Investments in tracking and tracing systems in the food Industry: A PLS analysis. Food Policy. 2012;**37**(1):102-113

[8] Kagermann H, Anderl R, Gausemeier J, Schuh G, Wahlster W. Industries 4.0 in a Global Context: Strategies for Cooperating with International Partners. New York: Herbert Utz Verlag. 2016. pp. 19-23

[9] Folinas D, Manikas I, Manos B. Traceability data management for food chains. British Food Journal. 2006;**108**(8):622-633

[10] Kagermann H. Change through digitization—Value creation in the age of Industry 4.0. In: Management of Permanent Change. Wiesbaden: Springer Gabler; 2015. pp. 23-45

[11] Otto B, ten Hompel M, Wrobel S. International data spaces. In: Digital Transformation. Berlin, Heidelberg: Springer Vieweg; 2019. pp. 109-128

[12] Lee J, Kao HA, Yang S. Service innovation and smart analytics for Industry 4.0 and big data environment. Procedia CIRP. 2014;**16**:3-8

[13] Zheng Z, Xie S, Dai H, Chen X, Wang H. An overview of blockchain technology: Architecture, consensus, and future trends. In: 2017 IEEE International Congress on Big Data (BigData Congress). IEEE; 2017, June. pp. 557-564

[14] Zyskind G, Nathan O, Pentland A. Enigma: Decentralized computation platform with guaranteed privacy. arXiv preprint arXiv:1506.03471; 2015

[15] Roos, J. Identity Management on the Blockchain. Chair of Network Architectures and Services, Department of Computer Science, Technische Universität München. 2018. p. 105. https://doi.org/10.2313/NET-2018-11-1_14

[16] Breivold HP, Sandström K. Internet of things for industrial automation--challenges and technical solutions. In: 2015 IEEE International Conference on Data Science and Data Intensive Systems. IEEE; 2015. pp. 532-539

[17] Jacobovitz O. Blockchain for Identity Management. The Lynne and William Frankel. Beer Sheva: Center for Computer Science Department of Computer Science. Ben-Gurion University; 2016

[18] Boudguiga A, Bouzerna N, Granboulan L, Olivereau A, Quesnel F, Roger A, et al. Towards better availability and accountability for iot updates by means of a blockchain. In: 2017 IEEE European Symposium on Security and Privacy Workshops (EuroS&PW). IEEE; 2017. pp. 50-58

[19] Rusi T, Lehto M. Cyber threats mega trends in cyber space. In: ICMLG 2017 5th International Conference on Management Leadership and Governance. Academic Conferences and Publishing Limited; 2017. p. 323

[20] Novo O. Blockchain meets IoT: An architecture for scalable access management in IoT. IEEE Internet of Things Journal. 2018;**5**(2):1184-1195

[21] Sharma PK, Chen MY, Park JH. A software defined fog node based distributed blockchain cloud architecture for IoT. IEEE Access. 2017;**6**:115-124

[22] Dorri A, Kanhere SS, Jurdak R. Towards an optimized blockchain for IoT. In: Proceedings of the Second International Conference on Internet-of-Things Design and Implementation. ACM; 2017, April. pp. 173-178

[23] Jacobs SB. The energy prosumer. Ecology Law Quarterly. 2016;**43**:519

[24] Nakamoto S. Bitcoin: A Peer-to-Peer Electronic Cash System. 2008. http://bitcoin.org/bitcoin.pdf

[25] Hargrave J, Sahdev N, Feldmeier O. How value is created in tokenized assets. In: Blockchain Economics: Implications of Distributed Ledgers-Markets, Communications Networks, and Algorithmic Reality. ICCS 2018. Cambridge MA. Vol. 1. 2018. p. 125

[26] Catalini C, Gans JS. Initial Coin Offerings and the Value of Crypto Tokens (No. w24418). Cambridge, MA: National Bureau of Economic Research. 2018

[27] Adhami S, Giudici G, Martinazzi S. Why do businesses go crypto? An empirical analysis of initial coin offerings. Journal of Economics and Business. 2018;**100**:64-75

[28] Feng C, Li N, Lu B, Wong MH, Zhang M. Initial Coin Offerings, Blockchain Technology, and White Paper Disclosures; 2018

[29] Hsieh YY, Vergne JP. Bitcoin and the rise of decentralized autonomous organizations. Journal of Organization Design. 2018;**7**(1):14

Chapter 8

Leveraging Blockchain for Sustainability and Open Innovation: A Cyber-Resilient Approach toward EU Green Deal and UN Sustainable Development Goals

Paula Fraga-Lamas and Tiago M. Fernández-Caramés

Abstract

In 2015, the United Nations (UN) member states identified 17 Sustainable Development Goals (SDGs) to be fulfilled by 2030. SDGs are an urgent global call for action to provide a blueprint for shared prosperity in a sustainable world. At a European level, in December 2019, the European Green Deal was presented, a roadmap to implement the UN 2030 agenda with a commitment to a growth strategy that will turn environmental challenges into opportunities across all policy areas. To achieve these SDGs, blockchain is one of the key enabling technologies that can help to create sustainable and secure solutions, since it is able to deliver accountability, transparency, traceability, and cyber-resilience, as well as to provide a higher operational efficiency in global partnerships. This chapter overviews the potential of blockchain to face sustainability challenges by describing several relevant applications. Finally, different open challenges and recommendations are enumerated with the aim of guiding all the stakeholders committed to the development of cyber-resilient and high-impact sustainable solutions.

Keywords: blockchain, distributed ledger technology (DLT), cybersecurity, smart contracts, global challenges, open innovation, sustainability, SDGs, circular economy, blockchain4SDGs

1. Introduction

In December 2019, the European Commission (EC) unveiled a plan to become the first climate-neutral organization worldwide by 2050. The so-called European Green Deal [1] is a roadmap for setting the sustainability and well-being of citizens at the center of policymaking and then turning climate and environmental challenges into opportunities across all policy areas. As it was created, the EU Green Deal is a commitment with sustainable development and a fundamental part of the

EC strategy to undertake the United Nations (UN) 2030 Agenda for Sustainable Development [2]. The 17 Sustainable Development Goals (SDGs) involve the three dimensions of sustainability (economic, social, and environmental) and require all the stakeholders to act in a global collaborative partnership. Such goals aim to achieve no poverty and hunger, to grant access to health services, to improve infrastructures, to reduce inequality, to fight climate change, to protect marine ecosystems, or to promote alliances between different actors to improve people's lives, among others.

Emerging technologies like the Internet of Things (IoT), 3D/4D printing, augmented reality/mixed reality/virtual reality (AR/MR/VR), cyber-physical systems (CPSs), robotics, novel human-machine interfaces (HMI), artificial intelligence (AI), big data techniques, machine learning (ML), deep learning (DL), 5G/6G connectivity, and new computing paradigms, when oriented toward SDGs, will bring a wide range of disruptive solutions in multiple fields. Nonetheless, the mentioned technologies will create ever-increasing complex systems in terms of heterogeneity, autonomy, interoperability, and scalability that will also come with additional cybersecurity risks and threats of malicious attacks.

Distributed ledger technology (DLT) represents nowadays an evolution toward the so-called Web 3.0, the Internet of Value. This new era of the Internet will include a collaborative economy among peers with crowdsourcing data sharing systems [3, 4]. A blockchain is a specific type of DLT that involves timestamped blocks of transactions linked in a chain by cryptographic hashes. Blockchain presents a decentralized architecture that provides benefits in terms of security, privacy, non-repudiation, integrity, accountability, transparency, robustness, and authentication. Moreover, it provides a high operational efficiency and eliminates the need for centralized parties and/or intermediaries. In fact, the World Economic Forum (WEF) forecasts that, by 2027, 10% of the global gross domestic product (GDP) will likely be stored on DLTs [5].

In this context, blockchain and other DLTs can enable global partnerships for open innovation and cyber-resilient applications compliant with the aims of the EU Green Deal and the UN SDGs. Thus, the contribution of this chapter is to provide a global overview of blockchain as an enabler for sustainability and open innovation. In addition, its aim is also to make the different involved stakeholders to rethink global development challenges to create cyber-resilient, decentralized, and high-impact sustainable developments.

The rest of the chapter is organized as follows. Section 2 overviews the basic concepts of blockchain. Sections 3 and 4 summarize the main principles of blockchain for sustainability and open innovation. Section 5 presents some relevant use cases of blockchain-based applications toward each of the SDGs. Section 6 summarizes the key main benefits of blockchain for SDGs and their main open challenges. Finally, Section 7 is devoted to conclusions.

2. Basic concepts of blockchain

A blockchain is a secured distributed ledger whose data are shared among peers [6–9]. In some blockchains like Bitcoin, decentralized miners validate every transaction (by following a consensus protocol), which allows them to solve the Byzantine Generals Problem (i.e., a situation where different parties must agree on a strategy and some of them may be corrupt, disseminate false information, or have intention to deceive). In the case of cryptocurrencies, the problem to be solved is called the double-spend problem: it must be guaranteed that the exchanged digital cash was not spent previously [6].

There are four main types of blockchains depending on who can access the stored data (private or public blockchains) and who can manage such data (permissionless or permissioned blockchains). Since a blockchain can store any kind of digital information, it could be the future of all secure transactions. Moreover, blockchain enables smart contracts, which consist of self-sufficient decentralized code that is executed autonomously according to a business logic. Furthermore, some blockchain platforms can also run decentralized applications, which are commonly called DApps [10].

Another important concept is the so-called decentralized autonomous organization (DAO), which can operate without requiring management hierarchy or a centralized authority [11]. The first DAO was launched in 2016 and raised $150 million worth of Ether (ETH) in 27 days. Nevertheless, DAOs are still very immature from the legal and security standpoints (e.g., a DAO attack due to code bugs led to a more than $50 million (ETH) theft in June 2016). Since 2016, a number of DAO initiatives have arisen (e.g., Steemit). In addition, the proliferation of DAOs is linked to the concept of decentralized autonomous society (DAS), in which citizens may be able to establish self-enforcing trade agreements without relying on centralized institutions of power and control.

It must be noted that a blockchain is not suited for every SDG-oriented application, which must fulfill the following main requirements:

- Trustworthy transactions are needed, but traditional databases do not cover the application needs.
- Data need to be updated by more than one stakeholder.
- There is a lack of trust among the entities that will update the data.
- The updaters are not willing to give the control of the database to a third party, and the involvement of intermediaries wants to be avoided when possible.
- A database could be used, but it is likely to be attacked (e.g., denial-of-service (DoS) attacks) or censored.
- Data redundancy in multiple distributed computers is needed.

Additional requirements could be involved, so several researchers have proposed more detailed decision frameworks about the use of blockchain [6, 12, 13].

It is worth mentioning that a detailed description of the different blockchain design aspects is out of the scope of this chapter, but the reader can find additional insights on the following recent works [4, 6, 8, 13].

3. Blockchain for sustainability

Sustainability is related to the effect that current actions will have upon the future. Such an effect can take many forms that vary depending on their nature, like the utilization of natural resources as a part of production processes, the waste management processes, the effects of competition among corporations in the same market, the enrichment of the community by creating employment, the produced pollution, the outbreak of a pandemic, or the relation with regulators. For example, if natural resources run out, then they may be no longer available (i.e., raw materials). Thus, the way in which economic, social, and environmental resources are efficiently managed is a key issue for long-term sustainability.

Recently, the EU has progressed significantly toward sustainability through the three main approaches [14]:

- Corporate social responsibility (CSR)/responsible business conduct (RBC) and new business models
- Business and human rights and the protection of human rights in general
- Sustainability and the implementation of the UN 2030 Agenda for Sustainable Development

The definition of CSR and RBC is related with ethical behavior and particularly with the relationship between a corporation and its stakeholders within a societal context, integrating social, environmental, and economic concerns into its business processes [14]. CSR/RBC can also be seen as actions under SDG 8 (decent work and economic growth).

In 2011, the UN Human Rights Council endorsed 31 Guiding Principles on Business and Human Rights (UNGPs) [14]. This approach came up as a sort of response to the perceived failure of CSR/RBC in terms of law binding and state oversight.

Recently, given the clear relationship between the three approaches (CSR/RBC, UNGPs, and the SDGs), the EC has adopted a holistic and practical approach toward sustainability irrespective of its name (i.e., CSR, RBC, business and human rights, SDG) while at the same time recognizing the target goal between the different agendas.

Within this context, blockchain is able to bring advantages toward sustainability in four main aspects: cybersecurity, accountability, transparency, and traceability:

- *Cybersecurity*. Applications for sustainability should be enabled by a robust digital infrastructure resilient to cyberattacks [15]. Cybersecurity should be implemented by design in the underlying technologies (e.g., IoT, AR, AI).
- *Accountability*. It is related to an organization (e.g., corporation or individual) acknowledgment of the impact of its actions, assuming responsibility for them. It implies to quantify the internal and external effects of the actions and report them to all the stakeholders. Such a reporting needs to be understandable, relevant, reliable, and comparable between different organizations and over time.
- *Transparency*. It implies that the external impact can be obtained from reporting by all the external stakeholders [16].
- *Traceability*. It is the ability to identify and trace assets (e.g., products, parts, processes, events, data, and materials) from their origins to production and distribution processes and, ultimately, until the end of their life cycle [17, 18]. Regarding Sustainable Supply Chain Management (SSCM), it also relates to human rights (e.g., fair trade, safety in labor, and privacy) and anti-corruption laws [18]. Therefore, it is a key organizational capability to foster sustainability. Two main categories can be considered within traceability [19]: internal (i.e., tracking and tracing assets within an organization) and external (i.e., it seeks to know the flow of information and assets between different logistics systems and processes among a number of organizations).

The importance of external traceability has been enhanced by globalization, the free movement of people and the global expansion of complex supply chain structures, combining networks of actors from multiple sectors (business, public, non-profit, and informal) in multiple locations.

4. Blockchain for open innovation

Open innovation, where innovative knowledge and ideas flow freely internally and externally to an organization, has become an important factor to enable sustainability [20]. To address SDGs, the EU recognizes the need for strengthening the impact of research and innovation and the use of coordinated approaches to ensure knowledge exchanges at an EU level [15]. These coordinated approaches will involve stakeholders with inter- and transdisciplinary points of view and the ability to manage jointly these development processes (SDG 17, partnerships for the goals) [21]. Although the current literature in open innovation details theoretical frameworks to guide solution development [20, 22], this development implies novel governance models that create thriving and diverse ecosystems where solutions are conceived, designed, experimented, implemented, supplied to the market, scaled up, and adopted. In that sense, one of the latest paradigms is called Open Innovation 2.0 (OI2) [23], a quadruple helix model where science, policy, industry, and society collaborate to achieve greater aims than a single entity.

Open innovation is uncertain and involves a high risk [20]. However, the lack of trust is today a major concern that withholds the cooperation and involvement of stakeholders in open innovation processes [24], especially for small- and medium-sized enterprises (SMEs). This need for orchestrating multiple stakeholders in a trusted and reliable way matches perfectly with the distributed nature of blockchain [20], which also provides the following main benefits:

- Stronger intellectual property (IP) protection. It includes responsible open-source licensing, processes of idea claiming [25], IP registries (e.g., trade secrets, patents, and trademarks), record keeping, licensing, and non-disclosure agreements (NDAs). In addition, profits (e.g., patent royalties and revenue on creative work) can be paid automatically according to predetermined agreements.

- Accurate collaboration between stakeholders modeled through smart contracts. Content can be shared among the stakeholders using smart contracts. Such smart contracts may deal with timestamping any IP disclosure or creation and automate corrective actions when unauthorized IP usage, IP infringements, and disclosure happen, acting as signed NDAs [25]. Furthermore, incentivized and rewarding mechanisms can be established (e.g., GlucoCoins to promote a global knowledge of diabetes [26]).

- Open data. It means the availability of data to all the stakeholders with a high degree of privacy (i.e., sovereignty and data ownership) and data protection.

- Regulatory compliance. It involves back-office processes mostly burdensome and inefficient to report to regulatory bodies. It also enables new open governance models.

5. Leveraging blockchain toward SDGs

Currently there are few examples of academic research on the use of blockchain for SDGs. For instance, the authors of [16] review recent academic and commercial "blockchain for good" applications in supply chain, innovations in governance, sharing economy, and financial inclusion. This section provides some relevant use cases of blockchain-based applications toward each of the SDGs. Such use cases are summarized in **Figure 1**.

5.1 SDG 1: no poverty

Access to credit and financial services (e.g., microfinance) is one of the most commonly known mechanisms to reduce poverty. For instance, crowdsourcing and crowdlending platforms can also ease financial inclusion. Blockchain can help to increase the efficiency, traceability, and transparency of these financial processes [27]. Moreover, micro-transactions and automatic funding through forecast-based financing [28] can be implemented jointly with smart contracts and big data analytics. Such models can provide more efficient funding, since no additional intermediaries are required and some procedures can be substantially simplified.

According to [29], 206.4 million people of 81 countries needed humanitarian assistance in 2018. For instance, only 6 of such countries represent 80.6 million people in need. Such a humanitarian assistance from governments and private donors reached US $28.9 billion in 2018. Nevertheless, a substantial percentage of the assistance was and is today lost due to fraud and corruption. Blockchain can be applied to provide tracking of the funds and to reduce cyberattacks. The authors of [28] highlight the need for ethical guidelines (i.e., privacy, intentional design choices, and humanitarian principles) and a common evaluation

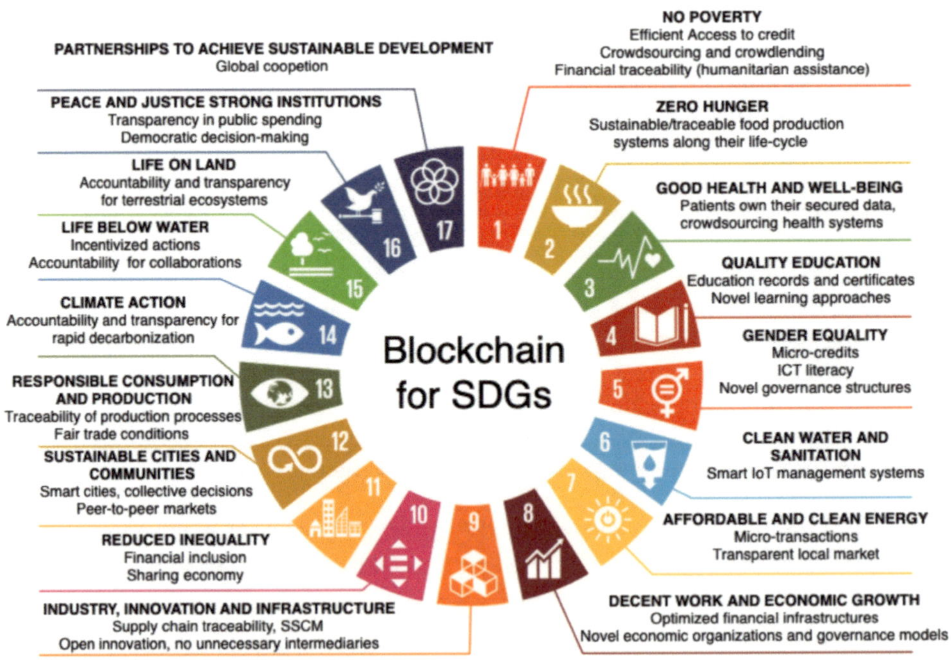

Figure 1.
Blockchain4SDGs: main blockchain use cases for SDGs.

framework of the solutions, especially as DLT developments are still in their early stages.

In 2017, the World Food Programme (WFP) [30] developed a proof of concept (PoC) in Sindh (Pakistan) named Building Blocks to evaluate blockchain for authentication and registration of transactions without financial intermediaries. Refugees have restrictions to open bank accounts and limited choices regarding the access and spending of their cash assistance. Building Blocks was also deployed with the aid of a biometric authentication system (i.e., iris scanning identification at checkout) in two refugee camps in Jordan to improve security and to ease cash transfers and the purchase of goods.

5.2 SDG 2: zero hunger

Sustainable food production systems along their life cycle can be guaranteed with the traceability properties of blockchain (e.g., avoid malpractice and guarantee food security).

5.3 SDG 3: good health and well-being

In Yue et al. [31], the authors propose a decentralized solution that enables healthcare intelligence that allows patients to control their data without compromising privacy or security.

In addition, blockchain can be used for managing data more efficiently during public health diseases. For instance, with the current rapid spread of the coronavirus disease (COVID-19) pandemic, a blockchain-based monitoring and traceability system can help to automatically identify unsafe areas by using geographic information and provide real-time information about patients (e.g., temperature, symptoms, and social distancing) for further analysis. As a result, it may keep communities from further infections and ensure (or even certify) that some locations (e.g., workplaces) are safe areas. For the implementation of such an application, cybersecurity and privacy (i.e., pseudo-anonymization) will be key issues for a successful deployment. Disease control may also depend on the ability of organizations (e.g., centers of disease control, state and local agencies, journalists, governments, hospitals, scientists) to collaborate in an effective and efficient manner. It must also be noted that richer countries are better prepared than poorer countries to identify a virus outbreak, to face infection with public health contingency plans, and to minimize the socioeconomical impact.

5.4 SDG 4: quality education

The authors of [32] have thoroughly reviewed the utilization of emerging technologies like blockchain, IoT, and fog and edge computing for improving education. Examples of applications include record verification [33], the management of digital copyright information [34], or the design and evaluation of novel learning approaches [35–37].

For instance, Sony Global Education [38] is an educational platform that uses Hyperledger Fabric to guarantee the authenticity of the student transcripts. Another commercial example is Learning Machine [39], a company that has created an open peer-to-peer infrastructure to issue digital records that can be easily shared and verified. The system is not only devoted to educational institutions: governments and companies can also issue blockchain-based records at scale, rooted in any blockchain they select.

5.5 SDG 5: gender equality

Easier access to financial services (e.g., even informal financial networks) promotes women empowerment as well as their independence. For example, hiveonline [40] is helping women through the CARE Village Savings and Loan Association (VSLA) program to get access to credits and markets with a fact-based reputation supported by blockchain. Such a financial infrastructure reduces the cost of cross-border payments and the risk of lending.

It must be noted that blockchain implies the use of Information and Communication Technologies (ICT), which can contribute to increase access to literacy. Furthermore, the inner characteristics of blockchain remove trust issues and enable the creation of new types of governance that may create equal opportunities for women leadership.

5.6 SDG 6: clean water and sanitation

Sustainable and efficient water management systems involve the use of sophisticated IoT architectures that optimize consumption and availability. Such architectures may be subject to security attacks (e.g., physical attacks on sensors, device cloning, data theft, DoS, jamming, or eavesdropping). Therefore, it is important to cyber-secure these systems and minimize the reliance on cloud-centered architectures that, when the server is down, may derive in the unavailability of the service. In addition, the communication between IoT devices within a decentralized architecture allows for avoiding single points of failure and enables the use of autonomous IoT transactions in a secure manner, thus guarantying tamper-proof data, visibility, and transparency in water trading [41].

5.7 SDG 7: affordable and clean energy

The authors of [42] study blockchain-based smart grid sustainable local energy markets. These systems enable cost-efficient micro-transactions, avoid central intermediaries, and promote reliability and equality among the different involved agents.

5.8 SDG 8: decent work and economic growth

Blockchain has the ability to promote economic growth by enabling free trade. For instance, it also has the potential to optimize global financial infrastructure in terms of asset transfer and operative costs.

In addition, it may ease new types of economic organization and governance (e.g., innovation-centered and governance-centered [43]). In Davidson et al. [43] the authors present an example of a self-governing organization for evaluating the contributions to projects on a network. When evaluating such an example, they introduce a wide range of perspectives to be considered, such as the problem of contractual enforcement, efficient institutions, governance, or even the constitutional characteristics of a nation.

5.9 SDG 9: industry, innovation, and infrastructure

The Industry 4.0 paradigm is expected to represent the next phase in the digitalization of all the sectors in the economy [8]. Supply chain traceability has been traditionally performed by wireless technologies like radio-frequency identification

(RFID) [44], which can be enhanced with additional security capabilities [45]. The next step forward is the so-called smart label [46], which adds novel features like event detection, interaction, and IoT capabilities. Such IoT solutions link cyber and physical worlds while enabling tracking and monitoring of assets and processes. Thus, blockchain goes one step further, making feasible end-to-end transparency in global supply chains. Business data can be shared rapidly between the different stakeholders across a trusted network [13]. In addition, smart contracts provide lower transaction costs by avoiding the intervention of intermediaries and third parties.

Ultimately, the ambition is to achieve Sustainable Supply Chain Management (SSCM), aiming to reduce the social and environmental impacts in global supply chains [47]. It is worth mentioning that, although research suggests that the combined use of blockchain and IoT devices will add significant value in supply chain, it will also impose some additional constraints in terms of computing power and power efficiency [6].

There are a number of supply chain projects deployed worldwide. For instance, Walmart, together with IBM, has developed a blockchain-based traceability system with Hyperledger Fabric [48]. In October 2016, they started with a PoC that tracked two items that were shipped to multiple stores. Before that, when a product had an issue (e.g., a customer became ill), it could take days to identify the batch, shipment, and vendor, and it may require to throw away a lot of the product. Through blockchain, it is possible to obtain specific data and details on the "how, where, and when" of the item within its supply chain. The shared database is able to capture attributes at the level of an individual package to take informed decisions. This functionality enables Walmart today to track a product in seconds (instead of days or sometimes weeks).

Following this approach, in August 2017, IBM announced a consortium with the food sector that included Walmart, Driscoll's, Dole, McCormick, Nestlé, Kroger, Tyson Foods, and Unilever. This consortium, named IBM Food Trust, will further explore the potential of blockchain to boost traceability along global supply chains with more products [48].

The shipping industry can also benefit greatly from blockchain. Ocean freight and maritime transport account for over 90% of the goods shipped globally [49]. The main characteristics are high number of involved stakeholders, complex transactions (e.g., letters of credit), burdensome paperwork, and lack of transparency, traceability, and information sharing. For instance, Maersk and IBM created TradeLens, a blockchain-based solution to create a more secure and efficient global logistics and spur industry-wide innovation [50].

Additionally, several blockchain startups are also innovating in traceability. For instance, the startup Provenance [51] has created an application to engage customers in the point of sale by providing mechanisms to verify sustainability claims (i.e., no greenwashing).

As it was previously mentioned, blockchain is also able to reduce transaction costs by reducing intermediaries and thus allowing more direct payment flows. For instance, DocuSign [52] is a company that offers several applications (e.g., electronic signature, contract lifecycle management). In 2015, DocuSign collaborated with Visa in a PoC project that used a smart contract to enhance car leasing processes. In 2018, DocuSign integrated an Ethereum blockchain in their signing services. As a result, the signers of an agreement can check anytime the integrity of the contract. DocuSign is also part of the Accord Project [53], a non-profit initiative that aims to develop a technology-agnostic ecosystem with open-source tools for smart contracts.

Accurate transaction records enable the use of tools for forecasting. For instance, Augur [54] is a decentralized platform built with Ethereum smart contracts that allows users to create their own prediction markets (i.e., oracle).

Another relevant commercial solution is Storj.io [55], which is a blockchain-enabled cloud storage network where users can rent the storage space that do not use and get paid in Storj tokens or store their information on a globally distributed network.

Some startups focus on removing intermediaries from trading like OpenBazaar [56], while other companies focus on providing visibility and transparency to philanthropy [57]. Such a global foundation leverages Bitcoin and blockchain to perform and track transactions while providing an immutable record of charitable financial transactions.

In 2018, IBM was awarded a patent for its Autonomous Decentralized Peer-to-Peer Telemetry (ADEPT) environment [58]. In 2016, IBM developed jointly with Samsung a PoC using different elements of Bitcoin to create a distributed network of IoT devices. For instance, to secure transactions, it uses a mix of proof of work (PoW) and proof of stake (PoS) as consensus protocols, BitTorrent for file sharing, Telehash for messaging, and Ethereum to support smart contracts.

It is also worth mentioning that other authors focused on smart grids and supply chain management systems as substantial areas of sustainable innovation [59].

5.10 SDG 10: reduced inequality

Theoretically, blockchain capabilities make the technology a catalyst for enabling a sharing economy with a democratic ownership structure (e.g., fractionally own goods by every community member) while avoiding unnecessary intermediaries. Nevertheless, authors like Novak [60] evaluate the implications of blockchain for income inequality and consider that, although it has potential to have a positive impact, it may also exacerbate current wealth concentration.

5.11 SDG 11: sustainable cities and communities

The authors of [61] propose a systematic literature review on specific blockchain use cases proposed by the research community. They remark the great concern about the infancy stage of blockchain.

5.12 SDG 12: responsible consumption and production

Production patterns can be monitored by using supply chain traceability techniques.

5.13 SDG 13: climate action

Blockchain will likely play an important role on the urgent actions for improving the accountability and transparency of policies to limit global fossil fuel consumption and foster decarbonization. Hyperledger, as part of the Linux Foundation, has recently announced a new Special Interest Group (SIG) that will explore how blockchain can help to address the climate goals set out in the Paris Agreement [62].

5.14 SDG 14: life below water

The company Possible Future oriented one of its projects to the sustainable use of the oceans, preserving their life and restoring damaged coral reefs [63]. They created a game, named CryptoCorals, in which for each purchase of a virtual coral, another coral is planted. The project is developed, thanks to the collaboration of a non-governmental organization (NGO) partner, and blockchain is used to guarantee transparency, as it is one of the major concerns of potential users.

5.15 SDG 15: life on land

Blockchain can be used to register trustworthy data about the different terrestrial ecosystems.

5.16 SDG 16: peace and justice strong institutions

Blockchain can help to reduce paper-based processes, minimize fraud, create inclusive institutions, and increase accountability in public services.

A good example is Delaware Blockchain Initiative [64], which was born with the aim of creating a legal framework for DLT sharing in corporations and governments. A more ambitious approach is Aragon [65], which is a startup that aims to create worldwide decentralized organizations, including employees and contractors from developing countries.

Other initiatives focus on increasing the transparency of democratic processes and on avoiding potential frauds. An example is Follow My Vote [66], which is a cost-effective online voting platform that audits ballots in real time.

5.17 SDG 17: partnerships to achieve sustainable development

To strengthen the means of collaboration between stakeholders is the key for enabling open innovation and for achieving SDGs.

6. Blockchain benefits and challenges toward SDGs and open innovation

The following paragraphs summarize the key main benefits that blockchain will bring to SDGs and their main open challenges.

Blockchain may provide significant operational benefits, since current information systems rely on centralized databases that operate in silos. By having a single, timestamped, immutable, and unique version of the truth, transparency and simplified audits can be guaranteed.

Furthermore, re-balancing the degree of information symmetry between stakeholders will help to achieve SDGs and will enable new forms of corporate governance and decentralized corporations. A collaborative mindset (the so-called coopetition) will be necessary to find additional ways to create value.

In terms of the maturity of the technology, there are a number of open challenges related to scalability, interoperability, standardization, or even energy consumption. The process of mining public networks, especially in the case of Bitcoin [67], requires enormous amounts of electricity. Therefore, although the underlying networks can provide sustainable applications, their footprint cannot be neglected [68].

From the cybersecurity standpoint, it is essential to provide secure applications with no single point of failure that comply with the expected degree of privacy. Nonetheless, it must be noted that blockchain can be also subject to cyberattacks [6]. The evolution of quantum computers will affect the security of public-key cryptosystems and hash functions. For instance, the authors of [9] analyze how to evolve blockchain cryptography to resist attacks based on Grover's and Shor's algorithms.

7. Conclusions

Blockchain can be used to develop secure peer-to-peer platforms for exchanging assets without intermediaries and in a trustworthy, sustainable, accountable, and transparent way to fulfill UN SDGs and the objectives of the EU Green Deal. Although research into blockchain has significantly increased in the last few years, there are not many academic or commercial solutions with sustainability and open innovation in mind. Moreover, most of them present solutions at very early stages of development.

Blockchain has the potential to radically change many societal sectors and to foster open innovation in all types of organizations, including supply chains, or the enforcement of governance in a completely innovative way. This overview has inherent methodological limitations due to its length and high level, so only a sample selection of some of the recent solutions is presented to give an idea of the potential of blockchain. The solutions described are not meant to be representative or generalizable. Such cases are the basics for further research, having in mind how blockchain can solve many of the current cybersecurity issues. Furthermore, open challenges were mentioned as a guidance for researchers and companies for future developments.

Abbreviations

IoT	Internet of Things
CPS	cyber-physical system
CSR	corporate social responsibility
DLT	distributed ledger technology
DoS	denial of service
PKI	public-key cryptography
NGO	non-governmental organization
PoC	proof of concept
P2P	peer-to-peer
PoS	proof of stake
PoW	proof of work
SSCM	sustainable supply chain management

Author details

Paula Fraga-Lamas[1,2*†] and Tiago M. Fernández-Caramés[1,2*]

1 Faculty of Computer Science, Department of Computer Engineering, Universidade da Coruña, A Coruña, Spain

2 CITIC Research Center, Universidade da Coruña, A Coruña, Spain

*Address all correspondence to: paula.fraga@udc.es and tiago.fernandez@udc.es

† The work of P. Fraga-Lamas was supported in part by BBVA and the British Spanish Society.

References

[1] Communication from the Commission to the European Parliament, the Council, the European Economic and Social Committee and the Committee of the Regions. The European Green Deal. 2019. Available from: https://ec.europa.eu/info/sites/info/files/european-green-deal-communication_en.pdf [Accessed: 20 March 2020]

[2] United Nation's 2030 Agenda and Sustainable Development Goals. Available from: https://sustainabledevelopment.un.org/ [Accessed: 20 March 2020]

[3] Tapscott D, Tapscott A. Blockchain Revolution: How the Technology behind Bitcoin Is Changing Money, Business, and the World. New York, NY, USA: Random House; 2016

[4] Fernández-Caramés TM, Fraga-Lamas P. Design of a fog computing, blockchain and IoT-based continuous glucose monitoring system for crowdsourcing mHealth. In: Proceedings of the 5th International Electronic Conference Sensors and Applications. 2018. p. 16. DOI: 10.3390/ecsa-5-05757

[5] World Economic Forum. Deep shift technology tipping points and societal impact. Survey Report. September 2015. Available from: http://www3.weforum.org/docs/WEF_GAC15_Technological_Tipping_Points_report_2015.pdf [Accessed: 20 March 2020]

[6] Fernández-Caramés TM, Fraga-Lamas P. A review on the use of blockchain for the internet of things. IEEE Access. 2018;**6**:32979-33001. DOI: 10.1109/ACCESS.2018.2842685

[7] Salman T, Zolanvari M, Erbad A, Jain R, Samaka M. Security services using blockchains: A state of the art survey. IEEE Communications Surveys & Tutorials. 2019;**21**(1):858-880. DOI: 10.1109/COMST.2018.2863956

[8] Fernández-Caramés TM, Fraga-Lamas P. A review on the application of blockchain to the next generation of cybersecure industry 4.0 smart factories. IEEE Access. 2019;7:45201-45218. DOI: 10.1109/ACCESS.2019.2908780

[9] Fernández-Caramés TM, Fraga-Lamas P. Towards post-quantum blockchain: A review on blockchain cryptography resistant to quantum computing attacks. IEEE Access. 2020;**8**: 21091-21116. DOI: 10.1109/ACCESS.2020.2968985

[10] Fernández-Caramés TM, Blanco-Novoa O, Froiz-Míguez I, Fraga-Lamas P. Towards an autonomous industry 4.0 warehouse: A UAV and blockchain-based system for inventory and traceability applications in big data-driven supply chain management. Sensors. 2019;**19**:2394. DOI: 10.3390/s19102394

[11] Wang S, Ding W, Li J, Yuan Y, Ouyang L, Wang F. Decentralized autonomous organizations: Concept, model, and applications. IEEE Transactions on Computational Social Systems. 2019;**6**(5):870-878. DOI: 10.1109/TCSS.2019.2938190

[12] Lo SK, Xu X, Chiam YK, Lu Q. Evaluating suitability of applying blockchain. In: Proceedings of the 22nd International Conference on Engineering of Complex Computer Systems (ICECCS), Fukuoka, Japan, 5–8 November. 2017. p. 158161

[13] Fraga-Lamas P, Fernández-Caramés TM. A review on blockchain technologies for an advanced and cyber-resilient automotive industry. IEEE Access. 2019;7:17578-17598. DOI: 10.1109/ACCESS.2019.2895302

[14] Commission Staff Working Document. Corporate Social Responsibility, Responsible Business Conduct, and Business & Human Rights: Overview of Progress. March 2019. Available from: https://ec.europa.eu/docsroom/documents/34963 [Accessed: 20 March 2020]

[15] Orientations towards the First Strategic Plan for Horizon Europe Revised following the Co-design Process. Version of 31 October 2019. Available from: https://ec.europa.eu/info/sites/info/files/research_and_innovation/strategy_on_research_and_innovation/documents/ec_rtd_he-orientations-towards-strategic-plan_102019.pdf [Accessed: 20 March 2020]

[16] Adams R, Kewell B, Parry G. Blockchain for good? Digital ledger technology and sustainable development goals. In: Handbook of Sustainability and Social Science Research. 2017. p. 127140. DOI: 10.1007/978-3-319-67122-2_7

[17] ISO 9000:2015(en). Quality management systems. Fundamentals and vocabulary. 2015. Available from: https://www.iso.org/obp/ui/#iso:std:iso:9000:ed-4:v1:en:term:3.6.13 [Accessed: 20 March 2020]

[18] United Nations. Global Compact (2014): A Guide to Traceability: A Practical Approach to Advance Sustainability in Global Supply Chains. Available from: https://www.unglobalcompact.org/docs/issues_doc/supply_chain/Traceability/Guide_to_Traceability.pdf [Accessed: 20 March 2020]

[19] Jeppsson A, Olsson O. Blockchains as a solution for traceability and transparency [master thesis]. Department of Design Sciences, Faculty of Engineering LTH, Lund University; 2017

[20] Yoon B, Shin J, Lee S. Open innovation projects in SMEs as an engine for sustainable growth. Sustainability. 2016;**8**(2):146. DOI: 10.3390/su8020146

[21] Rauter R, Globocnik D, Perl-Vorbach E, Baumgartner RJ. Open innovation and its effects on economic and sustainability innovation performance. Journal of Innovation & Knowledge. 2019;**4**(4):226-233. DOI: 10.1016/j.jik.2018.03.004

[22] Chesbrough H, Bogers M. Explicating open innovation: Clarifying an emerging paradigm for understanding innovation. In: Chesbrough H, Vanhaverbeke W, West J, editors. New Frontiers in Open Innovation. Oxford: Oxford University Press; 2014. pp. 3-28; Forthcoming

[23] Open Innovation 2.0 (OI2) web page. Available from: https://ec.europa.eu/digital-single-market/en/open-innovation-20 [Accessed: 20 March 2020]

[24] Gaggioli A, Eskendari S, Cipresso P, Lozza E. The middleman is dead, long live the middleman: The trust factor and the psycho-social implications of blockchain. Frontiers in Blockchain. 2019;**2**:20. DOI: 10.3389/fbloc.2019.00020

[25] De La Rosa JL, Torres-Padrosa V, El-Fakdi A, Gibovic D, Hornyák O, Maicher L, et al. A survey of blockchain technologies for open innovation. In: Proceedings of the 4th Annual World Open Innovation Conference. 2017. pp. 14-15

[26] Fernández-Caramés TM, Froiz-Míguez I, Blanco-Novoa O, Fraga-Lamas P. Enabling the internet of mobile crowdsourcing health things: A mobile fog computing, blockchain and IoT based continuous glucose monitoring system for diabetes mellitus research and care. Sensors. 2019;**19**: 3319. DOI: 10.3390/s19153319

[27] Kshetri N. Potential roles of blockchain in fighting poverty and

reducing financial exclusion in the global south. Journal of Global Information Technology Management. 2017;**20**(4):201-204. DOI: 1097198X.2017.1391370

[28] Zwitter A, Boisse-Despiaux M. Blockchain for humanitarian action and development aid. Journal of International Humanitarian Action. 2018;**3**(1):1-7. DOI: 10.1186/ s41018-018-0044-5

[29] Development Initiatives. Global Humanitarian Assistance Report 2019. Available from: https://devinit.org/ publications/global-humanitarian-assistance -report-2019/ [Accessed: 20 March 2020]

[30] Building Blocks. Blockchain for Zero Hunger. Available from: https:// innovation.wfp.org/project/building-b locks [Accessed: 20 March 2020]

[31] Yue X, Wang H, Jin D, Li M, Jiang W. Healthcare data gateways: Found healthcare intelligence on blockchain with novel privacy risk control. Journal of Medical Systems. 2016;**40**:218. DOI: 10.1007/s10916-016-0574-6

[32] Fernández-Caramés TM, Fraga-Lamas P. Towards next generation teaching, learning, and context-aware applications for higher education: A review on blockchain, IoT, fog and edge computing enabled smart campuses and universities. Applied Sciences. 2019; **9**(21):4479. DOI: 10.3390/app9214479

[33] Han M, Li Z, He JS, Wu D, Xie Y, Baba A. A novel blockchain-based education records verification solution. In: Proceedings of the 19th Annual SIG Conference on Information Technology Education, Fort Lauderdale, FL, USA, October. 2018

[34] Hori M, Ono S, Miyashita K, Kobayashi S, Miyahara H, Kita T, et al. Learning system based on decentralized learning model using blockchain and SNS. In: Proceedings of the 2018 10th International Conference on Computer Supported Education, Funchal, Portugal, 15–17 March. 2018

[35] Wu B, Li Y. Design of evaluation system for digital education operational skill competition based on blockchain. In: Proceedings of the IEEE 15th International Conference on e-Business Engineering (ICEBE), Xi'an, China, 12–14 October. 2018

[36] Lizcano D, Lara JA, White B, Aljawarneh S. Blockchain-based approach to create a model of trust in open and ubiquitous higher education. Journal of Computing in Higher Education. 2019;**32**:109-134. DOI: 10.1007/s12528-019-09209-y

[37] Zhong J, Xie H, Zou D, Chui DK. A blockchain model for word-learning systems. In: Proceedings of the 5th International Conference on Behavioral, Economic, and Socio-Cultural Computing, Kaohsiung, Taiwan, 12–14 November. 2018

[38] Sony Global Education official webpage. Available from: https://www. sonyged.com/ [Accessed: 20 March 2020]

[39] Learning Machine official webpage. Available from: http://www.learning machine.com/ [Accessed: 20 March 2020]

[40] Hiveonline official webpage. Available from: https://www.hivene twork.online/ [Accessed: 20 March 2020]

[41] Dogo EM, Salami AF, Nwulu NI, Aigbavboa CO. Blockchain and internet of things-based technologies for intelligent water management system. In: Artificial Intelligence in IoT. Cham: Springer; 2019. pp. 129-150

[42] Mengelkamp E, Notheisen B, Beer C, Dauer D, Weinhardt C. A blockchain-based smart grid: Towards

sustainable local energy markets. Computer Science-Research and Development. 2018;**33**(1–2):207-214. DOI: 10.1007/s00450-017-0360-9

[43] Davidson S, De Filippi P, Potts J. Economics of Blockchain. 2016. Available from: SSRN 2744751

[44] Fraga-Lamas P, Fernández-Caramés FM, Noceda-Davila D, Vilar-Montesinos M. RSS stabilization techniques for a real-time passive UHF RFID pipe monitoring system for smart shipyards. In: Proceedings of the IEEE International Conference on RFID (IEEE RFID); Phoenix, AZ, USA; May. 2017, p. 161166

[45] Fernández-Caramés TM, Fraga-Lamas P, Suárez-Albela M, Castedo L. A methodology for evaluating security in commercial RFID systems. In: Crepaldi PC, Pimenta TC, editors. Radio Frequency Identification. 1st ed. Rijeka, Croatia: IntechOpen; 2017

[46] Fernández-Caramés TM, Fraga-Lamas P. A review on human-centered IoT-connected smart labels for the industry 4.0. IEEE Access. 2018;**6**: 25939-25957. DOI: 10.1109/ ACCESS.2018.2833501

[47] Pagell M, Wu Z. Building a more complete theory of sustainable supply chain management using case studies of 10 exemplars. Journal of Supply Chain Management. 2009;**45**:3756. DOI: 10.1111/j.1745-493X.2009.03162.x

[48] Walmart case study. Available from: https://www.hyperledger.org/resource s/publications/walmart-case-study [Accessed: 20 March 2020]

[49] IMO (International Maritime Organization). Available from: https:// business.un.org/en/entities/13 [Accessed: 20 March 2020]

[50] Maersk and IBM Introduce TradeLens Blockchain Shipping Solution. Available from: https://ne wsroom.ibm.com/2018-08-09-Maersk-and-IBM-Introduce-TradeLens-Bloc kchain-Shipping-Solution [Accessed: 20 March 2020]

[51] Provenance official webpage. Available from: https://www.provena nce.org/ [Accessed: 20 March 2020]

[52] Docusign and Blockchain official web page. Available from: https://www. docusign.com/products/blockchain [Accessed: 20 March 2020]

[53] Accord project. Open source software tools for smart legal contracts. Available from: https://www.accord project.org/ [Accessed: 20 March 2020]

[54] Augur official webpage. Available from: https://augur.net/ [Accessed: 20 March 2020]

[55] Storj.io official webpage. Available from: https://storj.io/ [Accessed: 20 March 2020]

[56] OpenBazaar official webpage. Available from: https://openbazaar.org/ [Accessed: 20 March 2020]

[57] BitGive Foundation official webpage: Available from: https://www. bitgivefoundation.org/ [Accessed: 20 March 2020]

[58] Autonomous Decentralized Peer-to-Peer Telemetry (ADEPT) patent. Available from: https://patents.google. com/patent/US20170310747A1/en [Accessed: 20 March 2020]

[59] Lund EH, Jaccheri L, Li J, Cico O, Bai X. Blockchain and sustainability: A systematic mapping study. In: Proceedings of the 2019 IEEE/ACM 2nd International Workshop on Emerging Trends in Software Engineering for Blockchain (WETSEB), Montreal, QC, Canada. 2019. pp. 16-23. DOI: 10.1109/ WETSEB.2019.00009

[60] Novak, M. The implications of blockchain for income inequality.

Blockchain Economics: Implications of Distributed Ledgers-Markets, Communications Networks, and Algorithmic Reality. Vol. 1. 2019. p. 235. DOI: 10.1142/9781786346391_0012

[61] Shen C, Pena-Mora F. Blockchain for cities—A systematic literature review. IEEE Access. 2018;**6**: 76787-76819. DOI: 10.1109/ACCESS.2018.2880744

[62] Hyperledger Climate Action and Accounting SIG. Available from: https://wiki.hyperledger.org/display/CASIG/Climate+Action+and+Accounting+SIG+Home [Accessed: 20 March 2020]

[63] Possible future Cryptocorals. Available from: https://www.possible-future.com/project/cryptocorals-blockchain-and-the-oceans/ [Accessed: 20 March 2020]

[64] Delaware Blockchain Initiative official webpage. Available from: h ttps://corpgov.law.harvard.edu/2017/03/16/delaware-blockchain-initiative-transforming-the-foundational-infrastructure-of-corporate-finance/ [Accessed: 20 March 2020]

[65] Aragon official webpage. Available from: https://aragon.org/ [Accessed: 20 March 2020]

[66] FollowMyVote official webpage. Available from: https://followmyvote.com/ [Accessed: 20 March 2020]

[67] de Vries A. Bitcoins growing energy problem. Joule. 2018;**2**(5):801805. ISSN: 2542-4351. DOI: 10.1016/j.joule.2018.04.016

[68] Giungato P, Rana R, Tarabella A, Tricase C. Current trends in sustainability of bitcoins and related blockchain technology. Sustainability. 2017;**9**(12):2214. DOI: 10.3390/su9122214